T0252007

INSTRUCTOR'S MANUAL
to

Chris Park's
THE ENVIRONMENT

INSTRUCTOR'S MANUAL
to

Chris Park's
THE ENVIRONMENT

Greg Lewis

Routledge
Taylor & Francis Group

LONDON AND NEW YORK

First published 1998 by Routledge
2 Park Square, Milton Park, Abingdon, Oxon, OX14 4RN

Simultaneously published in the USA and Canada
by Routledge
711 Third Avenue, New York, NY 10017, USA

First issued in hardback 2017

Routledge is an imprint of the Taylor & Francis Group, an informa business

© 1998 Greg Lewis

Typeset in Galliard by Solidus (Bristol) Ltd, Bristol, England

All rights reserved. No part of this book may be reprinted or
reproduced or utilized in any form or by any electronic, mechanical,
or other means, now known or hereafter invented, including
photocopying and recording, or in any information storage or
retrieval system, without permission in writing from the publishers.

British Library Cataloguing in Publication Data
A catalogue record for this book is available from the British Library

Library of Congress Cataloging in Publication Data
A catalogue record for this book has been requested

Publisher's Note
The publisher has gone to great lengths to ensure the quality of this reprint
but points out that some imperfections in the original may be apparent

ISBN 13: 978-1-138-42445-6 (hbk)
ISBN 13: 978-0-415-16663-8 (pbk)

Contents

Figures

Acknowledgements

I would like to express thanks to Chris Park for entrusting me with the task of preparing this manual. Thanks also to Hugh Cutler for the Leopold matrix and the rest of the staff in the Geography Department at the University College of St Martin, Lancaster, for their support and illustration for the final section on alternative assessment. The burden of getting thoughts and scribblings onto disk was done efficiently by Sonia Barnes and Sarah Walker to whom I am extremely grateful. Finally, I wish to thank my family, for their support and tolerance during the many hours given to this work.

Greg Lewis
Senior Lecturer in Geography
University College of St Martin, Lancaster

Introduction

This manual complements Chris Park's textbook *The Environment*. The content and layout of *The Environment* enables a systematic and varied approach to its use as a course text, and the layout of this manual reflects this strength. The purpose of the manual is to facilitate the preparation of student learning experiences and assignments from *The Environment* by teachers.

The Environment presents the Earth as a system, underpinned by a hierarchical nest of natural systems interacting with each other. Within this framework the actions of humans are introduced and system outcomes explored. Major themes relating to our scientific knowledge of natural system operation, the linkages between systems and the scale (both spatial and temporal) of system interference and reaction are unifying emphases in this manual. The manual is divided into chapters that replicate the chapter headings in *The Environment* so that topical content can be accessed readily.

The manual considers each chapter of *The Environment* under the following headings, allowing a full use of the textbook resources plus additional complementary material:

Aims

This provides a list of the main thematic aims of the chapter. It may be used to emphasize the chapter's topical links as well as providing a unifying rationale for the material presented.

Key-point summary

This is a summary of the key-points in the main text. They are designed to reflect the main themes of the textbook with an emphasis on spatial aspects and timescales in relation to system workings. Applied knowledge is also emphasized.

Main learning hurdles

Students studying *The Environment* are likely to have a variety of academic backgrounds. This section highlights potential problem areas. The learning hurdle topics are addressed in general teaching terms relating to the concepts and techniques of approaching each topic rather than providing more detailed knowledge of the topic. This may be accessed from books in the main text section.

Key terms

These provide a quick reference guide to key terminology within each chapter. They allow the instructor to assess the terminology prior to teaching but may also be utilized as short informal summary definitional tests for the students. A non-assessed small group term and definition session would probably be most relevant. This could, possibly, be encouraged outside the normal timetabled sessions in student self-help groups.

Issues for group discussion

These should help instructors prepare structured seminar sessions. They aim to allow the student to enter wider environmental debate from a position of theoretical knowledge accrued from the appropriate chapter and relevant selected readings.

Selected reading

This provides a list of journal articles that appear in

each chapter of the text. They are from readily available journals reflecting a range of both length and presentation. Hopefully, this should allow flexibility in the construction of a discussion session dependent on ability and time.

Main texts

Additional information may be obtained from these to help the instructor widen the scope of each chapter. They may also be utilized to provide course and topical bibliographies for the students. The emphasis has been on a few recent publications though older key texts may also be listed. Some of Chris Park's further reading selection may also be reaffirmed in this personal, non-comprehensive choice.

Essay questions

A list of essay questions is provided using topics introduced in each chapter but encouraging further in-depth reading. The academic level and length of answer to these questions can be readily altered by interchanging the instructional words. As a range of sources is being encouraged, initial instruction of correct text and bibliographic referencing may be required to promote appropriate presentational standard and avoid plagiarism. The extensive use of referenced illustration in *The Environment* should be favourably commented on in relation to producing good environmental/geographical essays.

Multiple-choice questions

The multiple-choice questions provide a more formal medium for testing the students' 'factual' knowledge in relation to the key terminology, theories and their proponents plus the scale of events derived from both the text and figures/tables. This latter area is of great importance for the students' awareness of the magnitude of environmental operation, impacts and assessment. Whilst this section is more structured than key-term discussion it is not a satisfactory assessed item at this level. More usefully it provides the individual student with a non-assessed quantitative record allowing focused review and revision of these elements of the chapter. Correct answers are identified by an asterisk.

Figure questions

A small number of figures provide the basis of a number of questions. These are designed to test observational, interpretative and applied skills. They should also provide a general guide to how to frame other questions to many of the illustrations available in *The Environment*.

Short-answer questions

These short-answer questions are derived from the information boxes provided in each chapter by Chris Park. As such they provide a means of testing the key knowledge obtained from each chapter by the student. Answers to these questions should be no more than a paragraph long. Model answers follow each question.

Additional references

A short list of some more recent material is included to help fill the gap between textbook submission and the complication of this manual. They are drawn from essentially mainstream journals providing, in general, easily accessible information with which to update material or provide additional case studies. As with the selected readings, a variety of levels is catered for from the advanced college student to the undergraduate.

Web sites

Information about our Earth, and the environmental issues facing it, is constantly being updated. A consistent Internet address is included which provides a doorway to more specialized sites. Whilst students should be encouraged to delve into this vast array of information, this encouragement should be tempered with the warning to be selective and not to download vast amounts of superficial information under the pretext of good research.

Aims

- To provide an overview of the text necessarily incorporating chapter topics but more importantly stating the approach and central themes (as outlined below).

- To introduce the following approaches:
  a systems framework
  an interdisciplinary perspective
  a global perspective.

- To introduce the central themes of:
  the environment as a hazard and the variety of disaster outcomes for individuals/society
  the significance and scale of environmental problems have a global relationship
  the need for sustainable methods of usage of the Earth and its natural resources
  a widening concern for the environment amongst the general public as well as specialists (scientists and politicians).

N.B. the following sections may be used to illustrate and contextualize the themes/approach of *The Environment*. Equally they may be used towards the end of a course to synthesize the topical knowledge from individual chapters into a thematic framework.

Key-point summary

- The Earth is our *'life-support'* capsule (our environment). Environmental damage to the systems that comprise the Earth have potentially disastrous consequences to individuals, communities and society as a whole.

- The Earth is an interacting system comprising the *lithosphere, atmosphere, hydrosphere* and *biosphere.* Affecting one will eventually affect the others.

- The human species has always affected these systems. Through time the effects have increased due to increased consumption and technological change.

- The environment has both physical and social aspects. Inevitably conflicts of interest will arise in our use and abuse of the environment.

- Contemporary environmental problems are symptoms of underlying abuse of the environment by society. Here human behaviour has, and does, affect the environment in such a way that health, economic possessions and natural systems may be jeopardized.

- The environment and its systems are under pressure from increased resource use. Underpinning this is the exponential growth of the human population. Over-exploitation of natural resources for food production purposes has precipitated soil exhaustion, deforestation and desertification; all degrading the terrestrial system. Similar pressure on the marine food system has resulted in depleted fish stocks. This pressure is not uniform across the Earth, depending both on the population distribution of actual numbers and the consumption per capita of particular groups. This is the basis of the *inequality* of resource use.

- Environmental problems are symptoms of conflict between physical systems (the environment) and our social systems. In order to understand the root causes, as a step to the solving, of environmental problems a *multidisciplinary approach* is required. This incorporates methods and knowledge from both the natural and social sciences.

- Large areas of the land, air and oceans have no formal ownership. They are viewed as common resources. If open access to these *'commons'* is allowed then they become overused, e.g. common pasture land, pollutants emitted into the atmosphere, or common fishing grounds in the North Sea. There is no responsibility on anyone to limit their use of the 'commons'. The resulting situation is that people will try to increase their usage (economic benefit) of the common property resource. Degradation of the quality, and possible reduction in the quantity, of the common property resource will follow. This is the *Tragedy of the Commons*.

- The right to *sovereignty* has the potential to reverse this tragedy. Having a stake in both the costs and benefits allows a country to balance these, resulting in a cessation of overuse. However, *growth*, *internal competition* and *private ownership* may make this outcome problematic.

- There are a number of aspects that make contemporary environmental issues quite different from those of the past. Human influence now affects *global environmental systems*. The spatial *scale* of the major problems is felt world-wide, e.g. ozone layer depletion and global warming. Even point source pollution problems are *transboundary* in nature, e.g. Chernobyl fallout and acid rain. The rates of interference to environmental systems are greater (and increasing with time) than ever before, e.g. soil erosion and biodiversity loss. These effects on the environment are now seen to be *persistent,* e.g. radioactive waste and species extinction. This has ramifications for future generations. Ecological systems are approaching *thresholds*: their natural limits. With system breakdown we are threatening our life-support system and again leaving problems for future generations. Scientific knowledge about how the Earth's systems function and interact is imperfect. This *uncertainty* over total system functioning leads to problems with forecasting and predicting cause–effect relationships. Most importantly uncertainty means that there is often conflict over the evidence between interested parties. The summation of these aspects suggests that action is required now. How the environment is used at present will affect what type of environment is left in the future.

- A dichotomy of viewpoints exists between protagonists of *Limits to Growth* based on unsustainable resource use and *cornucopians* who believe societies' knowledge and technology will ensure resources are sustainable into the future. Importantly both use quantitative scientific research to back their arguments.

- The *Rio Earth Summit (1992)* was a culmination of international conferences that finally set an agenda for future planetary management. This was based on a holistic, sustainable development approach as opposed to the more discrete nature conservation programmes evident in the late 1970s to the early 1980s. Global system interactions were now formally recognized at an international political level.

- Recent environmental research has a major focus on using new technologies to monitor environmental systems. The need for *global monitoring* has necessitated a more intense and wider field of view, particularly the use of satellites as monitoring platforms. There has also been a move to standardizing techniques and data form so that local and regional environmental information can be harmonized to produce a larger, consistent picture of system interactions.

- The crisis facing the Earth and its human population is produced by detrimental change in environmental systems, most seriously at the global scale. These changes increase the scale and frequency of *environmental hazards*. The response from society has been a rise in environmentalism with at the general policy level a recognition of sustainable development as the future course for human interaction with the Earth's resources and systems.

Main learning hurdles

Themes and the human-environment system

As an overview chapter the main themes in the book are introduced. The instructor needs to emphasize that these themes need consideration regarding topical issues in the subsequent chapters. A key area that should be fully understood is the general interaction of the resource use system between humans and the natural environment, especially the pollution outcomes and feedback links.

Environmentalism

The historical roots of environmentalism are interesting counterpoints to the scientific theory and system knowledge prominent throughout *The Environment*. The key-point to get across is the wide variety of environmental organizations and the various shades of 'green principles'.

Limits to Growth

The instructor must explain the context of and evidence used in the Limits to Growth argument. Time scale and spatial nature of the evidence must be presented so that students can assess its validity. Similarly the evidence for the cornucopian technological argument must be examined. Ensure that the students are clear regarding the status and terminology of resource and reserve types as these are often inappropriately used in student work.

International environmental conferences

It is sometimes difficult for students to differentiate between the outcomes and terminology of the major international environmental meetings. The instructor should emphasize the major themes of the various conferences, the focus (specific or holistic) and most important the type of policy agreement and its legal implications, e.g. treaty, convention, protocol, resolution, recommendation or declaration.

Key terms

Agenda 21; Atmosphere; attitudes; baselines; biodiversity; biosphere; commons; Corine; cornucopians; ecological systems; economic growth; enclosure; environmental crisis; environmental problems; expectations; exponential growth; global change; Global Environment Facility; green; hydrosphere; IGBP; land-use changes; life-support system; Limits to Growth; lithosphere; monitoring and analysis; natural hazards and disasters; optimists; Our Common Future; persistence; population; remote sensing; Rio Earth Summit; sovereignty; sustainable yield; symptoms; thresholds; UNCED; uncertainty; UNEP; utopia; values; World Conservation Strategy.

Issues for group discussion

What are the problems of pollution monitoring and control?

Relating the general system diagram of pollution as waste output from the resource use system, the discussion should encapsulate the occurrence of pollution at a variety of scales (both spatial and temporal) and under numerous circumstances. General discussion regarding the rights to pollute (costs) and create material goods (benefits) should

ensue. The discussion should develop the global theme in relation to economically developed countries' attitudes and less economically developed countries' attitudes. Reference to the Brundtland Report would be useful.

How relevant is the Tragedy of the Commons?

The students should consider their own position, a regional and a national perspective before considering the common global picture. Discuss how difficult it is to change attitude and reach agreement. Consideration of areas such as the North Sea would provide a useful international focus on a range of resource issues both biotic and abiotic.

Are you an environmental pessimist or technological optimist?

The instructor should encourage the students to express their initial feelings at the start. They should continue by exploring the broad themes presented in the chapter. This should help the students frame their position from the start of the course. It is useful to record responses and positions so that at the end of the course, based on the accumulated course knowledge, a comparative discussion can take place to see if views have altered in response to the textbook material. Reference to both Groves (1992) and Degg (1992) would be useful to consolidate general viewpoints.

Selected readings

Degg, M. (1992) 'Natural disasters: Recent trends and future prospects'. *Geography* 77(3), 198–209.
A global view of the nature and distribution of natural hazards, with an emphasis on the past 50 years. The impact on countries at different levels of economic development is highlighted and human settlement distribution equated to future hazard scenarios.

Groves, R. H. (1992) 'Origins of Western environmentalism'. *Scientific American* 267(1), 22–7.
An historical view of the effects on the environment of economic activity. The role of scientific methods and knowledge is presented as underpinning strategies for nature conservation. The inertia of policy making in the light of scientific evidence before our contemporary environmental concern is clearly stated.

Textbooks

Adams, W. M. (1992) *Green Development: Environment and Sustainability in the Third World*. Routledge: London.
This is a clearly written book that links the principles of ecology to development issues. Key issues are approached topically and each is extremely well referenced.

Burton, I., Kates, R. W. and White, G. F. (1993) *The Environment as Hazard*, 2nd edn. Guildford: New York.
A readable account providing an overview of hazard migration policy. Extremely useful in clearly linking human response to changes in the atmospheric, hydrological and terrestrial systems. Wide-ranging set of topical examples covering most areas of the globe. Can be used to lead on to development issues and environmental response.

Chatterjee, P. and Finger, M. (1994) *The Earth Brokers*. Routledge: London.
An appraisal of the major actors in the process of producing environmental policy. Clearly illustrates the global scale of interaction and initiatives. This is presented in the context of multidisciplinary co-operation and conflict.

Cutter, S. L. (1993) *Living with Risk*. Edward Arnold: London.
Discusses technological hazards, including chemical pollution and nuclear waste in a detailed manner. The interaction of natural and social systems is clearly illustrated. A variety of spatial scales is used reinforcing the cascading nature of the Earth's systems.

Gore, A. (1992) *Earth in the Balance: Forging a New Common Purpose*. Earthscan: London.
A comprehensive discussion of environmental policy from a political perspective. Very readable account of the environmental crisis, possible action and future management.

Harper, S. (ed.) (1993) *The Greening of Rural Policy: International Perspectives*. Belhaven: London.
A series of essays based on a 'green' philosophy. Very useful accounts of national-scale effects on the global commons. A variety of case studies from the economically developed world is presented.

Harrison, P. (1992) *The Third Revolution: Population, Environment and a Sustainable World*. Penguin: London.
An affordable overview of the breadth of the environmental crisis. Clear, readable analysis. Useful topical chapters on biodiversity, marine pollution and atmospheric pollution.

Holdgate, M. (1995) *From Care to Action: Making a Sustainable World*. Earthscan: London.
Wide-ranging synthesis of the varied policy involved in the sustainability debate. Reflects in depth on the effects of development with particular relevance to biological systems.

McCormick, J. (1992) *The Global Environmental Movement*. Belhaven: London.
An examination of the history of environmentalism. It builds up a comprehensive picture of how the environment has become a global issue. This is illustrated with reference to both the economically developed and developing worlds. It may be used to pose many debating points for students to consider.

Middleton, N. (1995) *The Global Casino: An Introduction to Environmental Issues*. Edward Arnold: London.
This has comprehensive coverage of a wide range of key environmental issues. It is well illustrated and provides a sound introduction to both the working of the physical environment and the human systems (political, social and economic) that interface with it.

O'Riordan, T. (ed.) (1995) *Environmental Science for Environmental Management*. Longman: Harlow.
A comprehensive introduction to the interactions between the Earth and its inhabitants. It focuses on clear links between the natural world and human societies. This is presented within an interdisciplinary approach. Scientific knowledge about the Earth's systems are used to present scenarios for policy decision making and rational environmental management.

Pepper, D. (1996) *Modern Environmentalism: An Introduction*. Routledge: London.
An accessible text for students, providing a coherent introduction to the beliefs and ideas of environmentalists. The role of science in providing fuel for the environmental debate is a key theme, illustrating how scientific knowledge and theories of our environment are used to frame important issues.

Pickering, K. T. and Owen, L. A. (1997) *An Introduction to Global Environmental Issues*, 2nd edn. Routledge: London.
An up-to-date student-friendly text that provides an

extremely readable introduction to a comprehensive range of environmental issues. It is fully illustrated with many case studies providing stimulating material suitable for group discussion. The Earth's natural systems are explained concisely and at a relevant scientific level for college/undergraduate students. This material is well integrated into the consideration of environmental issues from the socio-economic, cultural and political points of view.

Roberts, N. (ed.) (1994) *The Changing Global Environment*. Blackwell: Oxford.
A good undergraduate introductory text illustrating the link between environmental system processes and the consequences of interference. The series of essays is presented in a coherent and integrated fashion highlighting some of the key global issues of the present. Many contrasting environments are discussed and illustrated at a variety of scales.

Simmons, I. G. (1993) *Environmental History: A Concise Introduction*. Blackwell: Oxford.
A clear introduction providing an outline of the history of human–environment interaction. It presents concepts and approaches from many disciplines in an understandable manner. The synergistic nature of the multi-disciplinary approach is emphasized along with an appreciation of the different scales of change instigated by human intervention.

Watts, S. and Halliwell, L. (eds) (1996) *Essential Environmental Science: Methods and Techniques*. Routledge: London.
A useful manual for students wishing to conduct primary research into environmental systems. Individual chapters focus on fieldwork techniques in surveying, sampling, soil and water systems. It usefully also includes social environmental research methods. Most appropriate in demonstrating the multidisciplinary range of techniques and information that can be used in the study of the environment.

Essay questions

1 Evaluate the view that our use of the Earth's resource systems is proceeding at a non-sustainable rate.
2 Discuss the issues that need to be considered by policy makers in giving priority to ecological sustainability over economic needs.
3 Is it possible to reconcile the variety of green environmental perspectives to provide a united case for sustainable development?
4 Are you optimistic or pessimistic about the future sustainability of the Earth?
5 'Environmental issues that are perceived to be global problems are essentially local or regional problems.' Discuss.
6 Outline why economically less developed countries may find it more difficult to conserve resources.
7 Explain the difference between environmental hazards and human disasters.
8 Describe the main factors that have increased environmental hazards through time.
9 How might remote sensing be used to aid sustainable development?
10 'Is the world overcrowded?' Evaluate in relation to sustainable development.

Multiple-choice questions

Choose the best answer for each of the following questions. The correct answers are asterisked.

1 How many main environmental systems is the Earth comprised of?
   (a) 1
   (b) 4   *
   (c) 7
   (d) 10

2 The number of people living in areas where air quality is bad for health is:
   (a) 625 million   *
   (b) 62 million
   (c) 6 million
   (d) 1 million

3 Effects of ozone increase in the lower atmosphere of the USA have:
   (a) decreased crop yields by 10 per cent   *
   (b) increased crop yields by 10 per cent
   (c) decreased crop yields by 25 per cent
   (d) had no effect on crop yields

4 Tropical forests cover:
   (a) 26 per cent of the world's land surface
   (b) 16 per cent of the world's land surface
   (c) 6 per cent of the world's land surface   *
   (d) 0.6 per cent of the world's land surface

5 What percentage of the Earth's wildlife species live in the tropical forests?
   (a) 29 per cent

(b) 59 per cent
(c) 90 per cent   *
(d) 99 per cent

6 Between 1958 and 1986 by how much did the global fish catch increase?
(a) stayed the same
(b) 2 times
(c) 3 times   *
(d) 5 times

7 How long did it take world population to increase from 5 to 6 billion?
(a) 1 year
(b) 11 years   *
(c) 100 years
(d) 1000 years

8 Manufacturers have been prosecuted for polluting the air in London since:
(a) the mid-twentieth century
(b) the late nineteenth century
(c) the late sixteenth century
(d) the early fourteenth century   *

9 Globally the loss of topsoil through wind and water erosion per day is:
(a) 2 hundred tonnes
(b) 2 thousand tonnes
(c) 2 million tonnes
(d) 200 million tonnes   *

10 Utopia was first described by:
(a) Thomas Cranmer
(b) Thomas More   *
(c) James Lovelock
(d) Thomas Malthus

11 The United Nations Conference on Environment and Development (UNCED) was widely referred to as the:
(a) Environment Summit
(b) Earth Summit   *
(c) Sustainable Summit
(d) Global Summit

12 UNEP stands for:
(a) United Nations Environment Programme *
(b) United Nations Ecological Programme
(c) United Nations Earth Priorities
(d) United Natural Environment Priorities

13 'Our Common Future' is often referred to as the:

(a) Environmental and Development Report
(b) Gro Harland Report
(c) Norwegian Report
(d) Brundtland Report   *

14 Agenda 21 is:
(a) a protocol
(b) an action programme   *
(c) a law
(d) a treaty

15 State of the environment reports comprise:
(a) an inventory of what is where
(b) an assessment of their state and quality
(c) a baseline against which to compare changes
(d) (a), (b) and (c)   *

16 IGBP stands for:
(a) International Geological-Biological Programme
(b) Inter-Governmental Baseline Project
(c) Inter-Governmental Biosphere Project
(d) International Geosphere-Biosphere Programme   *

17 The French satellite surveillance system is:
(a) GIS
(b) GOES
(c) NOAA
(d) SPOT   *

18 The International Decade for Natural Disaster Reduction started in:
(a) 1990   *
(b) 1980
(c) 1970
(d) 1960

19 Sustainable development involves:
(a) sustainable use of renewable resources
(b) biodiversity conservation
(c) minimal damage to natural environmental systems
(d) (a,) (b) and (c)   *

20 In the view of the Swedish Environmental Protection Agency producers should be liable for their goods:
(a) from cradle to grave   *
(b) during the production process
(c) after sale
(d) by adopting a wait and see policy

Figure questions

1 Figure 1.8 shows 'Limits to Growth' predictions for a range of important variables based on previous standard trends. Answer the following questions.
- (a) Describe the trends in the five curves during the next century forecasted by the model.
- (b) List the problems arising from these trends.
- (c) How would the cornucopian optimists view change this model?

Answer

(a) Resources will initially fall steeply until mid-century then follow a less steep path approaching equilibrium by the end of the century. All other variables initially rise before they peak and eventually fall, with the exception of food per capita which begins to recover in the last two decades. Both industrial output per capita and food per capita have similar upward growth trends and peak at about

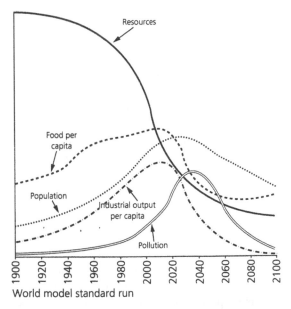

Figure 1.8 Limits to growth *predictions of global populations, resources and pollution. A number of different simulations were made by the MIT team, using the dynamic computer model (Figure 1.7), based on different assumptions. The one summarised here, referred to as the standard run, assumed that declining resources and increasing degradation of the environment would eventually result in declines in human aspects of the model. After Figure 1.8a in Cunningham, W.P. and B.W. Saigo (1992)* Environmental science: a global concern. *Wm. C. Brown Publishers, Dubuque*

2015. Both initially fall rapidly, until 2040, with industrial output continuing to decline towards zero by the end of the century. Population growth is similar to industrial output, peaking approximately 20 years later. The reduction in population during the rest of the century is even and less steep than either food per capita or industrial output per capita. Pollution shows both the steepest rise and also the longest lag to the peak, 2035. Pollution decline approximately mirrors the slope of growth and tends to just above zero by the end of the century.

(b) Resource depletion means a reduction in material goods and lower standard of living. A human economic problem rather than an environmental one. Also it might instigate 'resource wars'. Increased pollution will degrade natural systems with associated health effects likely to have continued effects even after the decline in output. The reduction in food per capita will lead to malnutrition and decline in population after the trend lines intersect. Industrial output per capita decline will again reduce the material standard of living, affect the global economic system and further reduce employment.

(c) Resources would be rising rather than falling due to more efficient usage of resources, substitution and recycling technology. This would counteract the falling trends in the other variables. Pollution would be problematic to the environment and may be the limiting factor.

2 Figure 1.12 shows the interactive relationship between people and the environment. Explain how the feedback loops affect environmental quality.

Figure 1.12 *Two-way relationship between people and environment. The relationship between people and environment is symbiotic, involving both resources (opportunities) and hazards (constraints). After Figure 2 in Park, C.C. (1991)* Environmental hazards. *Macmillan Education, London.*

Answer

Unless sustainably managed, resource use will deplete the environment. Usually this is manifested in loss of land (habitat) and a degraded landscape. The main threat to environmental quality via technology is from pollution. This is often less locationally specific than the abstraction of resources, though it relates to resource use via the manufacturing process. Technological pollution may be a by-product of the function, e.g. fossil fuel power station or car usage. It may also be caused by accidents with technology, e.g. Chernobyl, or a change in status of material by technology, e.g. nuclear waste or plastics.

Short-answer questions

1   What were the main triggers for the rising interest in the environment?

Answer

A growing awareness of the number of human activities that affect the environment. That the environment provides our life-support system on which we ultimately depend. The environment provides us with the natural resources which fuel societies' development. An increasing trend in the number of environmental hazards occurring through time. Visible evidence of degraded landscapes due to human actions. Increasing environmental lobbying suggesting we are approaching system breakdown. All these are underpinned by greater media coverage and more sophisticated, all-embracing global information networks.

2   How is population pressure spatially defined?

Answer

Globally, absolute numbers are important with exponential population growth outstripping the Earth's resources. Related problems at this scale are likely to occur in the future. At present it is the distribution of population in relation to accessing resources that causes problems. Locally and regionally, population densities exceed the carrying capacity of the resource base available. At the national/international scale population pressure manifests itself by the consumption per person.

Thus, there is a basic imbalance between the high-consuming North and low-consuming South.

3   What are the major causes of the environmental crisis?

Answer

The global environmental crisis is a product of societies' abilities, organization and attitudes. Technological ability has increased the rate at which we use resources. Greater numbers of people are able to be supported, putting increased pressure on the environment in particular locations of concentrated population. The global free market economic system has put economic expansion before environmental costs. This is compounded by the view of the major natural environmental system as 'free goods'.

4   What is the Tragedy of the Commons?

Answer

The degradation of natural common resources because individual interests override collective sustainable management. Efficient use of a common resource requires collective agreement regarding levels of use so that carrying capacity is not exceeded and the resource remains sustainable.

5   What was the most common forecast predicted by the Limits to Growth computer model?

Answer

Limits to Growth predicted large-scale system breakdown within the next century. Shortages of natural resources would lead to falling industrial growth, restricted food supplies and a population decline resulting from pollution, famine, disease and stress.

6   What are state of the environment reports?

Answer

They are up-to-date inventories of information on environmental quality and natural resource status. They allow a baseline assessment of environmental attributes and allow comparison of these through time so that changes may be monitored and recorded.

7 Outline significant developments in environmental monitoring during the 1990s.

Answer

During the 1990s there has been a concentration on global coverage with a harmonization of data collection between countries. Continuous monitoring has replaced time sampling with the increased use of computerized data handling systems to cope with the increased amount of data generated.

8 What are the benefits of using a global perspective in studying the environment?

Answer

The global perspective illustrates how widespread many environmental problems are. It illustrates the environmental pathways between places spatially disparate, via cause–effect relationships. Global problems impacting back on us at the local and regional scale clearly indicate our reliance on the global life-support system.

9 What key issues must sustainable development address?

Answer

Sustainable development must take into account both time and space. Through time the environmental heritage must be preserved for future generations. Ethically we should not expect them to deal with problems created by us. Contemporary issues relate to changing the emphasis from short-term economic gain to ensuring long-term environmental stability. Spatially the uneven access to resources needs addressing by more equitable distribution. Common resources also require sensible allocation amongst claimants.

10 How might you express environmental concern?

Answer

An individual may join a non-governmental environmental organization such as Greenpeace.

More proactively they may become involved in environmental politics or local environmental projects. As a consumer a more environment-friendly lifestyle may be adopted in relation to family size, transport and purchase of goods.

Additional references

Fell, N. (1996) 'Outcasts from Eden'. *New Scientist* 151 (2045), 24–7.
An account of the enormous problem of refugees fleeing a variety of environmental hazards. The interference of people on natural systems and the human cost are clearly discussed.

Pearce, F. (1996). 'Trouble bubbles for hydro-power'. *New Scientist* 150 (2028), 28–31.
Illustrates the uncertainties regarding the future due to the complexities of the Earth's systems. Unforeseen side effects of renewable energy systems can be more damaging than the polluting energy systems they replace. This is considered in relation to carbon dioxide and methane emissions and global warming.

Shcherbank, Y. M. (1996) 'Ten years of the Chernobyl era'. *Scientific American* 274(4), 32–7.
An appraisal of the lasting environmental and health effects of the Chernobyl nuclear disaster. Both local and more widespread areas are examined. A final summary of the total economic cost and damage to the nuclear industry is included.

Smith, K. (1996) 'Natural disasters: definitions, databases and dilemmas'. *Geography Review* 10(1), 9–12.
Clearly defines and quantifies the variety of hazards. These are illustrated at both the global and selected national scales. A variety of discussion points are raised.

Web site

www.ulb.ac.be/ceese/sustvl.
An extremely comprehensive catalogue of a wide variety of environmental topics. Useful for contacting organizations and dynamic links to major search engines.

Aims

- To provide an overview of systems and systems theory.

- To explore the structure and operation of environmental systems.

- To examine flows in biogeochemical cycles and relate these within the systems framework.

- To illustrate the application of the systems approach using real world case studies and exploring the structure and operation of the relevant systems.

Key-point summary

- The *River Nile* case study is utilized as an example of a *drainage basin system*. The management of such a system is *complex*. This case study illustrates the general nature of systems and the interaction between natural and human systems including:

1 natural systems provide multiple resources and these are used in a variety of ways by the people dependent on the system;

2 these resource uses often conflict with each other and are exacerbated by the layering of human system patterns over the natural system. In this case the demarcation of parts of the River Nile drainage basin by national boundaries;

3 the network pattern of the River Nile channel sub-system illustrates the various flow contributions to the total system. This interconnectivity presents the holistic nature of natural systems;

4 through time, with increasing population growth, humankind increases its reliance on natural systems. In order to limit the threat to this reliance society attempts to control natural systems. This exercise of control (*management*) may also be used proactively to attempt to maximize the resource;

5 as each part of a system has a link, either directly or indirectly, to other parts of the system; change in one part will impact on other parts of the system. In order to optimize the system's resources an integrated (holistic) approach to management is needed. This requires knowledge of the workings of the system

6 imperfect knowledge of natural systems' functioning means that management of systems will often produce unwanted outcomes (*problems*) as well as the intended results (*benefits*). In the management of the River Nile, via the Aswan High Dam Scheme, the benefits were increased control over river flow (*flood and drought control*), improved *irrigation* and the production of power (*hydroelectricity*). Unwanted environmental and social impacts were water loss through *evaporation* and *seepage*, *salinization*, displacement of people due to the land required for the scheme, loss of cultural heritage, seismic stress, disruption of water circulation adjacent to the Nile delta, lake infilling, channel and delta erosion, reduction in flood plain fertility, reduced fish catches and an increase in water-borne disease.

- Management of natural systems should be sympathetic to the systems' workings. Reduction in the pressure from human systems is more *sustainable* (*long term*) than short-term, hard engineering solutions (*techno fix*).

- Human activities and physical regions rarely have common boundaries. This spatial mis-match causes conflict in the use and allocation of natural resources, their control and management.

- A system is made up of component parts. Each component is *dynamically linked* within the system to other components. The complexity and detail at which the system may be analysed depends on the scale of study.

- The parts of the system work together to provide a functional system. A system has a purpose, e.g. the natural river system is a transfer system for water, sediment and nutrients.

- Systems have boundaries. This defines their relationship to other systems. An *isolated* system has no exchange of *energy* or *material* across its boundaries. A *closed* system can exchange energy but not material. An *open* system exchanges both energy and material freely across its boundaries.

- All environmental systems on Earth are inter-related and act as open systems with each other.

- As systems can be studied at a variety of scales and at different levels of detail, a hierarchical structure is employed for descriptive purposes:

  1 a *system* is the total environmental system in operation, e.g. a drainage basin;

  2 a *sub-system* is a major sub-division of the system. Often discreet and exhibiting system workings such as the flow of energy and material, e.g. hillslopes, floodplain and river channels within a drainage basin;

  3 system *components* are specific properties associated with the system or sub-system, e.g. the angle of slope of a hillside, or the amount of sediment carried by a river.

- The Earth functions as an *integrated system* comprising four main sub-systems: the *lithosphere*; the *atmosphere*, the *hydrosphere* and the *biosphere*. These open environmental systems have four key structural parts: *inputs* of material and energy, *outputs* of material and energy, *flows* by which material and energy move within the system and *stores* where material and energy remain for various time periods before being released back as flows.

- The operation of an environmental system depends on how the structural elements fit together. *Energy* drives the system. Effectively, it is in infinite supply (solar energy). The different types of energy are *chemical*, *electrical*, *heat*, *kinetic* and *potential*. The use of energy within an environmental system is governed by the *laws of thermodynamics*. Not all the energy available to

the system is used for work. Heat is dispersed waste energy. The waste energy is referred to as *entropy*. Naturally functioning systems maintain themselves with low entropy. Human activities involving energy usage often have higher entropy and therefore contribute to an increase in global entropy.

- In contrast materials within a system are finite but can be recycled. The three main types of material that flow through environmental systems are *water*, *nutrients* and *sediment*. Water is the integrating material that links the Earth's environmental systems.

- The dynamics of an environmental system may be understood in relation to energy flows:

  1 energy inputted into a system is not necessarily outputted in the same form;

  2 inputs and outputs of energy and materials do not have to balance in the short term as they may be stored in the system. Long-term balancing of inputs and outputs provides system stability. The integrity of stores, as well as flows, is a key aspect of maintaining this stability;

  3 in order to function environmental systems need a constant flow of energy into the system;

  4 environmental systems use energy inefficiently. Thus, the flow of energy does not achieve near its maximum potential for work within the system.

- Changes in a system component will often induce changes throughout the system. The system operates a *feedback* mechanism (sequence of events) which impacts back to the original system component that changed. *Positive* feedback enhances the change and leads to a breakdown of the system's stability (rare mechanism in environmental systems). *Negative* feedback dampens down the change providing self-regulation of the system.

- Most environmental systems are in *equilibrium*. Inputs and outputs to and from the system may be constant over the *long term* (*steady state*) or display a uniform change (*dynamic*). Note that these are long-term trends, and short-term oscillations occur about these general trends. As long as these short-term oscillations remain at a scale which does not fundamentally affect system behaviour, the system will remain stable. The speed of recovery from a sudden short-term change is known as the *resilience*. If the scale of change is of a magnitude that changes the system's behaviour, then the system *threshold* (critical point) will have been breached. This will lead to a new equilibrium state. Different systems have different time-response rates to change. This time delay is known as a *lag*.

- In applying systems theory and knowledge to environmental problems a holistic approach is advocated. The system is viewed in a *synergistic* fashion with the whole system being more than the sum of its component parts. We should, therefore, think of protecting the entire global system rather than concentrating on specific areas. An appreciation of the synthesizing nature of the biosphere and that it is a dynamic entity is crucial to this appreciation. The role of humans is a fundamental part of this global system.

- Despite our understanding of how systems work there are still uncertainties when employing this approach. Thus, future trends and responses to human actions are extremely difficult to predict.

- Using systems theory at the global scale clearly indicates the holistic nature of the biosphere's function, the integration of humanity within this global system and the dynamism of the total system to smaller-scale change. However, at this scale the interactive complexity of system processes means that there is still scientific uncertainty regarding how the biosphere will respond to change and at what rate.

- Chemical elements and their flows throughout the environment are key factors in the life-bearing capacity of the biosphere. The flow is cyclical between organisms and their environment (*biogeochemical cycles*). Human activities may influence these natural cycles by altering inputs, extracting, and altering the speed and direction of processes within the chemical system.

- The main biogeochemical cycles are of nitrogen, carbon and sulphur. Whilst there are still uncertainties about the operation of part of these cycles there is evidence of human-induced changes. The nitrogen cycle is affected by the addition of fertilizers and nitrogen-loaded air pollutants. Carbon and sulphur are added to the atmosphere by the burning of fossil fuels and natural stores of carbon reduced by vegetation removal.

- Biogeochemical cycles interrelate with the other major global environmental systems. Therefore, disturbance in these cycles triggers changes in the other environmental systems.

Main learning hurdles

Integrated management

Students tend to compartmentalize topics and in many ways simple systems thinking tends to foster this. It is important at an early stage to develop interrelationships between systems. The Nile water-management case study is a good example of the consideration of a number of systems, often considered from a multidisciplinary perspective. Thus, the instructor should highlight how a variety of Earth science disciplines combines with engineering, politics and economics to provide an integrated management policy.

Types of system

Many students have difficulty in unravelling complex system diagrams. It is useful to build up the complexity by using the black box (no system knowledge other than inputs and outputs), through the grey box (limited internal system knowledge) to the white box system where all internal workings are known. This also allows the instructor to demonstrate that we do not always have perfect knowledge of system operation and most environmental decisions are based on grey box examples. Definitions in relation to systems hierarchy also cause problems. A review of the section on system scale in this chapter is useful.

System operation

As systems are such an integrating theme throughout *The Environment*, time should be spent on system terminology and especially feedback mechanisms. A complex feedback loop often causes problems when viewed in entirety with a mixture of positive and negative connections.

It is useful to disaggregate the total feedback loop to produce sequential pairs of linked components viewed initially in isolation so that change in one will produce a predictable response in the other. When reassembled the initial trigger will have logically been worked through the system components to produce the final adjustment. As well as feedback, equilibrium, thresholds and time lags should be reviewed.

Biogeochemical cycles

Inevitably when pursuing further reading on biogeochemical cycles, especially in relation to human effects, chemical equations will be encountered. The instructor should ensure that students with a limited chemistry/biology background are not deterred. The main chemical formulae and reactions should be summarized in concert with the appropriate sections in this chapter.

Key terms

Biogeochemcial cycles; carbon cycle; complexity; conflicts; drought; energy budget; entropy; equilibrium; erosion; feedback; flooding; flows; gestalt; ground water; hydroelectricity; hydropolitics; inputs; irrigation; isolated, closed and open systems; lags; lake infilling; laws of thermodynamics; loss of nutrients; multinational; nested hierarchy; nitrogen cycle; osmosis; outputs; photosynthesis; resilience; salinization; seasonal storage; sub-system; sulphur cycle; synergy; system boundaries; system dynamics; system scale; techno-fix; thresholds; timescale; water management.

Issues for group discussion

Discuss the benefits and problems of the Aswan Dam scheme

As well as reviewing the case study in the chapter the students must read Pearce (1994) and Smith (1991). Quite diverse views should be expressed though the components of the Nile drainage basin must be explored in relation to both system feedback and the application of integrated management.

Discuss the importance of a global view of the environment

The students should be encouraged to explore spatial scale. A useful way is to build up a picture of the components in major biogeochemical cycles from local to global scale. A number of the selected readings emphasize large-scale interactions between the main environmental systems. A human perspective may be brought in by considering if, and by what mechanisms, the world is shrinking in human terms, e.g. communications.

Discuss the likely impacts of climate change

It would be useful for students to read Goudie (1993) and Park (1991). The students should focus their dis-

cussion on an evaluation of the scientific evidence and uncertainty in managing large-scale system change.

Selected reading

Goudie, A. (1993) 'Environmental Uncertainty'. *Geography* 78 (2), 137–41.
Considers the limits of knowledge about how the Earth's major systems function. Biodiversity, atmospheric systems and climate change are considered in the light of current scientific knowledge. A discussion of the complexity of natural systems is thematic throughout the paper.

Murray, J. W., Berber, R. T., Roman, M. R., Bacon, M. P. and Feely, R. A. (1994) 'Physical and biological controls on carbon cycling in the equatorial Pacific'. *Science* 266 (5182), 58–65.
Links system dynamics to biochemical cycling. The sequence and scale of the cycling mechanism is related to other atmospheric and oceanic system events, e.g. El Niño. A wide source of data is presented graphically.

Nriagu, J. O. and Pacyna, J. M. (1988) 'Quantitative assessment of world-wide contamination of air, water and soils by trace metals'. *Nature* 333 (6169), 134–9.
A global-scale view of human activities on natural environmental systems. Illustrates system feedback with contaminated system stores affecting human health.

Park, C. C. (1991) 'Trans-frontier air pollution: Some geographical issues'. *Geography* 76 (1), 21–35.
A variety of air pollution types are examined. The scientific evidence, public perception and management policy outcomes are all discussed. This clearly illustrates the complexity of effectively coping with human interference in large-scale environmental systems with different spatial responses.

Pearce, F. (1994) 'High and dry in Aswan'. *New Scientist* 142 (924), 28–32.
Demonstrates the impact of natural systems back to human influence. Usefully discusses the spatial implication of a natural system that transcends political boundaries.

Siegenthaler, U. and Sarmiento, J. L. (1993) 'Atmospheric carbon dioxide and the ocean'. *Nature* 365 (6442), 119–25.
The oceanic system store for atmospheric carbon dioxide is seen as crucial in the accurate prediction

of future global warming. System knowledge is required if sustainable decisions are to be made in controlling carbon dioxide emissions.

Smith, G. (1991) 'The Aswan Dam'. *Geography Review* 5(2), 35–41.
A positive appraisal of the Aswan Dam scheme in relation to Egypt.

Textbooks

Abu-Zeid, M. A. and Biswas, A. K. (1996) *River Basin Planning and Management*. Oxford University Press: Oxford.
Details the River Nile case study and considers future management and conflict at the sub-continental scale including North Africa and the Middle East.

Beard, J. M. (1994) *Chemistry, Energy and the Environment*. Wuerz: Winnipeg, Canada.
A readable introductory text that presents physics and chemistry theory in an understandable manner. There are particularly useful chapters on energy in the environment and biogeochemical cycles. The application of this theory is discussed in relation to a wide range of environmental pollutants.

Briggs, D. J., Smithson, P., Addison, K. and Atkinson, K. (1997) *Fundamentals of the Physical Environment*, 2nd edn. Routledge: London.
A comprehensive, chapter by chapter explanation of the operation of the Earth's major systems. Complex interactions are accessibly explained with extensive use of diagrams and boxed case studies.

Goudie, A. and Viles, H. (1997) *The Earth Transformed: An Introduction to Human Impacts on the Environment*. Blackwell: Oxford.
A very useful discussional text presenting a wide range of diverse case studies. These are thematically presented in relation to human impact on each of the major environmental systems.

Hugget, R. J. (1993) *Modelling the Human Impact on Nature: Systems Analysis of Environmental Problems*. Oxford University Press: Oxford.
An advanced text that clearly demonstrates the use of systems methodology in representing and evaluating impacts on our environment. Key model studies relating to biogeochemical cycles, climate change and the hydrological cycle are investigated.

Hugget, R. J. (1995) *Geoecology: An Evolutionary Approach*. Routledge: London.
The Earth's systems are defined as geo-ecosystems with changes examined within them, at their interface with each other, and in response to cosmic influence. The complexity and network of interdependencies at each of these scales are presented in a dynamic time framework.

Jakeman, A. J., Beck, M. B. and McAleer, M. J. (eds) (1993) *Modelling Change in Environmental Systems*. Wiley: Chichester.
An advanced text that illustrates how systems knowledge can be applied to measure effects on systems. It contains many case study chapters considering a variety of topical issues at a wide range of scales.

Newson, M. D. (1992) *Land, Water and Development: River Basin Systems and their Sustainable Management*. Routledge: London.
Examines the terrestrial environment at the drainage basin scale. Systems theory is used as a basis for an in-depth analysis of human–environmental interactions. A wide range of case studies explores management of these systems in countries of different economic status.

Nisbet, E. G. (1991) *Leaving Eden: To Protect and Manage the Earth*. Cambridge University Press: Cambridge.
A wide-ranging overview of the natural environment, how it functions and the effects of humanity. The first part of the book focuses on chemical and physical processes within a systems framework. Disruption of natural systems is then discussed using this terminology. A wide range of topical chapters on pollutants and energy sources are included.

Smil, V. (1997) *Cycles of Life: Civilization and the Biosphere*. W H Freeman: Basingstoke.
The main biogeochemcial cycles are examined with particular emphasis on human interaction. A logical progression from natural cycle operation, through human interference to natural system change is followed in a clear style.

Strahler, A. H. and Strahler, A. N. (1994) *Introducing Physical Geography*. Wiley: London.
A useful overview of the Earth. Discrete chapters on each of the major environmental systems allow an easy introduction for the student.

White, I. D., Mottershead, D. N. and Harrison, S. J. (1984) *Environmental Systems: An Introductory Text*. Allen and Unwin: London.

A comprehensive treatment of natural environmental systems suitable for first-year undergraduate study. All scales of system, and their components, are dealt with in a full and comprehensible manner.

Wilcock, D. (1983) *Physical Geography: Flows, Cycles, Systems and Change*. Blackie: Glasgow.
System definitions and applications are clearly presented to the student in an easily absorbed fashion. Good pre-reading text.

Essay questions

1 Illustrate and explain changes that would take place in the nutrient cycle following the clearing of natural vegetation and its replacement by cultivation at the local scale.
2 How might the knowledge of energy flows through environmental systems aid our management of them?
3 Discuss the concept of equilibrium over various timescales. Illustrate your answer with reference to physical systems.
4 Using examples from environmental systems explain both positive and negative feedback.
5 Discuss the main ways in which people can modify the nitrogen cycle.
6 Compare and contrast the transfer of nitrogen and sulphur between stores in their respective cycles.
7 Discuss the importance of scientifically understanding biogeochemical cycles.
8 Evaluate the impacts that alter the natural carbon cycle.
9 Explain the major stores and flows within the River Nile drainage basin.
10 Discuss the importance of time when studying environmental systems.

Multiple-choice questions

Choose the best answer for each of the following questions.

1 How much of Africa does the River Nile drain?
  (a) 25 per cent
  (b) 20 per cent
  (c) 10 per cent   *
  (d) 1 per cent

2 The record of yearly summer flow variation in the River Nile shows effects of:
  (a) Little Ice Ages
  (b) Roman settlement
  (c) cataracts
  (d) El Niño events   *

3 What stimulated the expansion of Egypt's cultivated area during the nineteenth century?
  (a) population growth   *
  (b) increased flooding of the Nile
  (c) rise in food prices
  (d) seasonal storage

4 The original Aswan Dam was built in:
  (a) 1902   *
  (b) 1912
  (c) 1934
  (d) 1960

5 The storage capacity of Lake Nasser is equivalent to:
  (a) 2 weeks' flow of the Nile
  (b) 2 months' flow of the Nile
  (c) 2 years' flow of the Nile   *
  (d) 2 decades' flow of the Nile

6 Which was estimated to be the largest user of water from the Aswan High Dam project?
  (a) municipal
  (b) industrial
  (c) irrigation   *
  (d) evaporation

7 Major development projects require consideration of long-term:
  (a) environmental impacts
  (b) costs
  (c) benefits
  (d) (a), (b) and (c)   *

8 A system that exchanges no energy or material across its boundaries is known as:
  (a) isolated   *
  (b) closed
  (c) open
  (d) integrated

9 The basic structure of an open environmental system has how many key parts?
  (a) 2
  (b) 4   *
  (c) 3
  (d) 5

10 The amount of original energy available to a system that is dispersed as heat is known as:

(a) exothermal
(b) environmental
(c) entropy   *
(d) radiation

11 'In a system of constant mass, energy cannot be created or destroyed, it can only be transformed' is:
(a) Newton's Law
(b) the first law of thermodynamics   *
(c) the second law of thermodynamics
(d) Boyle's Law

12 The typical residence time for smoke particles in the lower atmosphere is:
(a) a few weeks   *
(b) a few days
(c) a few hours
(d) a few months

13 The ability of a system to recover from a sudden change is known as its:
(a) fortitude
(b) resilience   *
(c) elasticity
(d) dynamism

14 System change that crosses a threshold promotes:
(a) a trigger
(b) continued equilibrium
(c) a new equilibrium   *
(d) (a), (b) and (c)

15 A lag is:
(a) an average state
(b) an initial equilibrium
(c) a delayed adjustment   *
(d) a trigger

16 Which of the following do plants not require large amounts of:
(a) oxygen
(b) carbon
(c) iron   *
(d) hydrogen

17 Elements are released from the organic phase of biogeochemical cycles by:
(a) combustion and decomposition   *
(b) combustion
(c) decomposition
(d) neither combustion nor decomposition

18 How much soil nitrogen is available to plants:
(a) 10 per cent

(b) 50 per cent
(c) 90 per cent   *
(d) 100 per cent

19 Global warming is mainly related to the:
(a) sulphur cycle
(b) nitrogen cycle
(c) methane cycle
(d) carbon cycle   *

20 Sulphur plays a key role in which of the following environmental problems:
(a) acid rain   *
(b) eutrophication
(c) traffic pollution
(d) global warming

Figure questions

1   Figure 2.4 shows the changes in the annual flow of the River Nile at Aswan (1945–88) and the variations in water stored in Lake Nasser (1965–90). Answer the following questions.
(a) What has been the effect of the Aswan High Dam on the annual flow at Aswan.
(b) Describe and account for the storage trends and variations in Lake Nasser.
(c) Explain the storage areas, indicated by dotted lines, of Lake Nasser.

Answers

(a) The Aswan High Dam has greatly decreased annual flow so that it is consistently below the long-term mean. There is still variability between years but the post-dam mean is much lower, suppressing flood flows.

(b) Initially storage water built up quickly over the first twelve years as the reservoir filled behind the Aswan High Dam. This was accomplished by restricted flood flow downstream as illustrated in the upper figure from 1966. From 1977 to 1981 a consistently high annual storage was achieved as annual demand was balanced by input into the reservoir. After 1981, to the low in 1988, reduced input from the upper catchment due to drought conditions restricted water supply. The latter two years reflected high flood conditions. Within this longer term trend there are seasonal fluctuations reflecting the flood season (high storage) and the demand for irrigation (low storage).

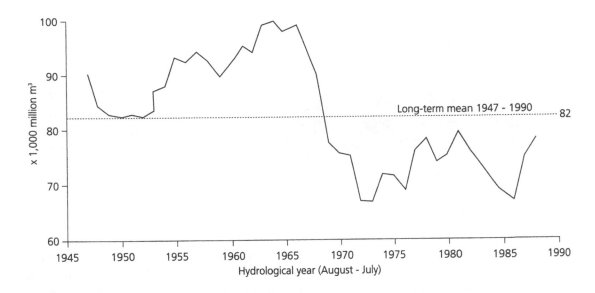

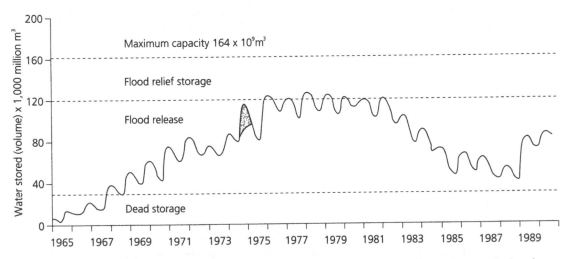

Figure 2.4 Changes in annual flow of the River Nile at Aswan, 1945–1988. The upper figure shows variations in natural flow (the volume before withdrawal of irrigation water) between 1945 and 1990, based on five-year running averages. The lower figure shows variations in the amount of water stored in Lake Nasser between 1965 and 1990. After Figures 3 and 5 in Smith, G. (1991) The Aswan Dam. Geography Review 5 (2); 35–41.

(c) The dead storage area is non-contributory water to the flow downstream. It is below the level of the turbines and does not contribute to power generation. Between the dead storage and flood relief storage is the normal storage which allows controlled flow to meet water demand and provide hydroelectric power. The flood relief storage is excess storage which can store abnormally high flood flow above the dam to be released gradually to reduce downstream system impact. This represents a safety storage capacity of 25% of total reservoir storage.

2   Figure 2.10 illustrates a forest as an open system. Answer the following questions.

(a) Describe and relate to other systems the inputs into the forest system.
(b) Describe and relate to other systems the outputs from the forest system.
(c) What are the major stores of water in the forest system.

Answers

(a) Solar energy is inputted from the cosmic system after passing through and being filtered by the atmospheric system. Precipitation is the input of water (in more than one state) from the lower atmospheric

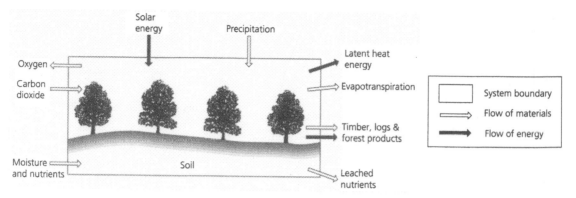

Figure 2.10 *A forest as an open system. A forest, like other ecosystems, displays, similar system properties to the water cycle (see Figure 2.9)*

system. Its input may be dependent on the topography of the local surface system or regionally in relation to the land–sea distribution related to the lithosphere. Carbon dioxide is inputted from the atmosphere being a product of the biosphere, both from natural and anthropogenic sources. Moisture and nutrients are inputted from adjacent soil systems as part of the hydrological cycle. Nutrients may also be inputted in precipitation.

(b) Energy is released from the system as latent heat energy; the waste product of a change of phase, e.g. evaporation of liquid to the atmosphere. It is also outputted as stored energy in forest products to the human economic system. Water is outputted from the forest system via surface evaporation and transpiration from the vegetational component. Vegetation also exchanges oxygen. Both materials transfer to the atmospheric system. Water and nutrients are outputted from within the soil component via hydrological pathways to adjacent land or rivers. Forest products are removed from the system to the human economic system.

(c) The major stores of water are the vegetation and the soils. Internally water flows from the soil to vegetation. Externally water may reach the soil store via vegetation.

Short-answer questions

1 What has caused salinization of the Nile floodplain?

Answer

Initially the Aswan High Dam has eliminated the natural flushing of mineral salts from the floodplain soils by controlling flood flow. Compounding this is the higher salinity, due to concentration by evaporation, of the new source of irrigation water from Lake Nasser. The greater surface water area of the irrigation canal network increases this salinity level further, again due to evaporation. The area is also susceptible to the updraw of groundwater, via capillary action as a product of the hot climate, with subsequent evaporation and concentration of mineral salts in the soil.

2 Describe the main differences between isolated, closed and open systems.

Answer

Isolated systems have no exchange of energy or material across their boundaries. Closed systems can exchange energy but not material across their boundaries, e.g. the global water cycle. Open systems can exchange both energy and materials across their boundaries, e.g. the drainage basin system.

3 Outline the Earth's four main environmental systems.

Answer

The four main environmental systems are the lithosphere, atmosphere, hydrosphere and biosphere. The lithosphere is the upper part of the body of the Earth. It comprises the crust and upper mantle and is made up of rocks and minerals. The atmosphere is the layer of air surrounding the Earth's surface. The hydrosphere contains all the surface and near surface water of the Earth. The biosphere is the region of the Earth in which life exists.

4 What is feedback?

Answer

Feedback occurs when a change in one system component produces a sequence of changes in other components forming a loop back to the original component. Feedback may be either negative, so that overall system state remains unchanged, or positive where a net change in the system state occurs.

5 What are the advantages of a systems approach?

Answer

The systems approach offers a holistic perspective of the environment. Its focus on interrelationships and synergistic processes fosters a broad understanding encouraging interdisciplinary co-operation. Practically it provides a framework for studying, describing, analysing and modelling the environment with a view to managing environmental problems.

6 Define biogeochemistry.

Answer

Biogeochemistry is the study of the cyclical exchange of elements between biotic and abiotic components of the biosphere.

7 What is the biogeochemical impact of burning fossil fuels?

Answer

Burning fossil fuels releases carbon into the atmosphere from the geological sediment store. The human increase in the use of these energy resources has increased the rate of this transfer of carbon tremendously. This atmospheric carbon, in the form of carbon dioxide, affects climate by enhancing the greenhouse effect. Overall the imbalance between carbon held in the surface system and that held in the atmospheric system will through time detrimentally affect, via global warming, the whole biosphere and its natural ecosystems via major environmental system change.

8 Outline the importance of nitrogen to life on Earth.

Answer

Fundamentally nitrogen is a constituent of biotic tissue. It is an essential gaseous part of the synergistic atmospheric mix (air). It is used by plants for growth and controls photosynthesis rates in many ecosystems. Nitrogen compounds are used within the human system for food manufacture and as fertilizers for agricultural systems.

9 Describe the basic forms of carbon found in the natural environment.

Answer

Carbon occurs in four basic forms. In its pure state it occurs as a mineral, e.g. diamond. It occurs as calcium carbonate found mainly in carbonaceous rocks, e.g. limestones. Within the atmosphere it is found as carbon dioxide. Carbon also occurs in the fossil fuel energy source as part of hydrocarbons.

10 What is the importance of biodegradability?

Answer

Biodegradability is the decomposition of waste materials, within biogeochemical cycles, by biological means. This provides for one flow of materials through natural systems. If waste material is not broken down it accumulates and therefore takes up useful space in the system. Sometimes biodegradability may be equated to the production of non-toxic residues.

Additional references

Lewin, R. (1996) 'All for one, one for all'. *New Scientist* 152 (2060), 28–33.
A reappraisal of the Gaia hypothesis in the light of modern complexity theory. Clearly portrays the notion of the 'superorganism' as the interaction of a large complex system.

Web site

www.met.inpe.br/geochem/home
A Brazilian site specializing in information on biogeochemical cycles. Much useful material on research at the land surface–atmosphere interface is discussed.

Aims

- To re-emphasize the holistic, self-contained nature of the Earth.

- To contextualize the Earth and its environmental systems within a much larger spatial and temporal scale. Especially regarding energy flows.

- To realize that our life-support system is fragile and, in the context of empirical knowledge, unique.

Key-point summary

- The energy flow from the Sun is the primary driving force for the Earth's environmental systems. In effect the Earth is a *solar-powered system*.

- We may look at the Universe and our position in it, i.e. as a small spaceship in the vastness of the cosmos, from a variety of perspectives. These may be scientific or teleological. The position we adopt will shape the way we view our planet and how we *'frame'* environmental issues.

- Human life is a result of, and preserved by, carefully controlled conditions. The Earth's energy relationship with the Sun and the environmental systems powered by this produce these critical conditions.

- The atmosphere is the most important system regarding the maintenance of human life. It also plays an important role in the other major environmental systems by control over *biogeo-chemical* and *water cycles*. This produces a unique climate appropriate to supporting higher species.

- A holistic way of viewing the Earth is provided by the *Gaia hypothesis*. This global representation of the environment as a self-regulating organism integrates all environmental processes and places human society in this context.

- As part of the *solar system* the Earth exists in an ordered system. Earth's range of motions within this system provides a patterned relationship to other bodies, providing cycles that affect the biosphere over a range of timescales. *Rotation* defines the length of day. *Revolution* around the Sun defines the seasons.

Main learning hurdles

The Gaia hypothesis and scale

Lovelock's Gaia hypothesis needs special emphasis as an environmental model. Most students will have an idea of science as analytical, reductionist and mechanical. They must be clear that Gaia portrays a world that is highly integrated, dynamic and adaptable. This last feature may cause some conflict with the idea of sustainability regarding the human species. As the Earth is explained as a powerful self-regulating system then we need not worry about sustainability. However, we are dependent on cycles (water and carbon) within the global system. Disruption of these could wipe out the human species without causing irreparable damage to the Earth. System change and adaption should be clearly explained in relation to the human timescale.

Daisyworld, radiation and feedback

System feedback may cause problems for some students. The negative and positive links between individual system components need exploring separately before the whole feedback loop is explored. The regulating system concept of Daisyworld benefits from this approach. The change in radiation properties associated with this model need explanation from the instructor, i.e. shorter wavelength incoming solar radiation being re-radiated as heat (longer wavelength infrared radiation) by black daisies and reflected unchanged by white daisies. Albedo as a regulatory mechanism requires emphasis.

Key terms

Carbon dioxide; equinox; Gaia hypothesis; galactic rotation; geophysiology; greenhouse effect; latitude; longitude; Lovelock; methane; nested hierarchy systems; precession; revolution; rotation; solstice; Spaceship Earth; symbiotic; thermostat.

Issues for group discussion

Discuss the relative evidence for dating the Earth

With illustration from Figure 3.4 students should be encouraged to frame the answer from a number of viewpoints; whilst the variety of scientific methods should be compared in relation to evidence from the major environmental systems. A useful exercise (that may require tactful handling) is the reconciliation of cultural viewpoints in relation to presented scientific evidence. At the end of the discussion the students should have a grasp of the logic behind each type of evidence and its accuracy. A subsidiary outcome is that the students are aware of the magnitudes of the time-scales discussed (Box 3.13).

Discuss the relative importance of physical, chemical and biological factors in the light of the Gaia hypothesis

With reference to Myers (1990) and Caldeira and Kasting (1992) the basics of the Gaia hypothesis should be presented. Discussion of the elements as a basis of life may be developed from Nullet (1994). The key outcome should be that after exploring the various inputs from the biogeochemical cycles, and the systems they link, students should begin to get a notion of synergy.

Discuss the importance of Gaia to the way we manage the Earth

Following on from the above discussion, Speier (1989) expands on Lovelock's term 'geophysiology'.

Discussion should be developed along two broad themes. First, the spatial patterning and division of the Earth into interdependent ecosystems. Second, the opportunity for multidisciplinary input into environmental management. This may link to the discussion in issue one, where a range of team skills and knowledge is required to unravel the complexity of the world. A key-point is that the students should be aware of the term's analysis and synthesis and how they relate to one another.

Selected reading

Caldeira, K. and Kasting J. F. (1992) 'The life span of the biosphere revisited'. *Nature* 360 (6406), 721–3.
Lovelock's predictions for the longevity of the biosphere based on the Gaia hypothesis are examined. Overall biosphere control from the solar-energy source is modelled in relation to life-giving systems on Earth. The conclusion is that these systems can function naturally much longer than Lovelock predicts.

Huggett, R. (1990) 'The bombarded Earth'. *Geography* 75(2), 114–27.
Explores the recent scientific knowledge accrued from space technology in relation to cosmic objects striking the Earth. This input from larger systems into the global system illustrates effects on major environmental systems. The likely scale of effect on the lithosphere, atmosphere and biosphere is related to the size and energy of such strikes.

Myers, N. (1990) 'Gaia: The lady becomes ever more acceptable'. *Geography Review* 3(3), 3–5.
A concise outline that clearly expresses the Gaia concept and how it may be applied to management. Uses some useful scientific evidence and explains the Daisyworld model succinctly.

Nullet, D. (1994) 'Unique Earth'. *Geography* 79 (1), 77–8.
A short paper that illustrates the complex interaction of the Earth's major systems. These underpin our unique life-support system in comparison to other bodies in the solar system. Provides a focus for discussion of what environmental variables are required to instigate and then support life.

Speier, R. (1989) 'The emergence of Gaia'. *Geographical Magazine* 61 (12), 30–3.
Emphasizes the importance of the Gaia hypothesis

for future planetary management. Readable account of major ecosystem interactions with multidisciplinary message.

Textbooks

Bak, P. (1997) *How Nature Works*. Oxford University Press: Oxford.
Written in mainly non-technical terms for the general reader this book critically looks at a wide variety of systems. Particularly relevant when exploring evolution and landscape formation. Complex systems are explained regarding their ability for self-organization.

Cattermole, P. (1995) *Earth and Other Planets: Geology and Space Research*. Oxford University Press: Oxford.
Contextualizes the unique 'life-support' nature of the Earth. Evidence supports areas of commonality within the larger context of the solar system. Key theories are introduced in student-friendly fashion and a non-jargon-ridden way.

Kauffman, S. (1996) *At Home in the Universe: A Search for Laws of Complexity*. Penguin: Harmondsworth
Clearly illustrates the biological complexity of ecosystems. The organization of the world and its species are explored at a variety of scales.

Lovelock, J. E. (1979) *Gaia: A New Look at Life on Earth*. Oxford University Press: Oxford.
Provides a holistic view of the Earth, integrating geological and biological systems with atmospheric composition. Its evolutionary approach clearly emphasizes the use of scientific evidence to build up environmental argument.

Press, F. and Seiver, R. (1993) *Understanding Earth*. W H Freeman: Oxford.
An introduction to the main fundamentals of geology. Useful illustrations explain functions and multidisciplinary thought is fostered in a discussion of applications.

Silk, J. (1994) *A Short History of the Universe*. W H Freeman: Oxford.
A concise description of current knowledge about the history and construction of the Universe. Radiation fluctuations and their implications are explored along with an appraisal of various theories of cosmology.

Essay questions

1 Why is the Earth unique in our solar system?
2 Discuss the theories explaining the origin of the Universe.
3 Discuss the importance of equating the Earth to a spaceship.
4 Compare and contrast the Pythagorean and Copernican views of the Universe.
5 Using a systems model illustrate the Earth as part of the cosmic system.
6 Evaluate the main components of our solar system.
7 Describe and account for the main motions of the Earth.
8 'Environmental systems may be viewed as operating like mechanical machines.' Discuss.
9 How might the Gaia model explain global climate control?
10 Evaluate the position of the human species in relation to the Gaia notion of how the Earth functions.

Multiple-choice questions

Choose the best answer for each of the following questions.

1 The Earth is essentially:
(a) an open system
(b) a closed system   *
(c) an isolated system
(d) a non-system

2 The approximate age of the Earth is:
(a) 4,600 million years   *
(b) 460 million years
(c) 46 million years
(d) 460 thousand years

3 The oldest dated rocks on Earth are:
(a) 3,800 million years old   *
(b) 380 million years old
(c) 38 million years old
(d) 380 thousand years old

4 The oldest dated rocks on Earth are in:
(a) Mexico
(b) France
(c) Greenland   *
(d) Denmark

5 What minimum diameter of asteroid could

potentially kill a large fraction of the world's population?
- (a) 200 km
- (b) 20 km
- (c) 2 km   *
- (d) 0.2 km

6 Longitude is based on:
- (a) latitude
- (b) the equator
- (c) meridians   *
- (d) oranges

7 Humans first travelled around the Moon in:
- (a) 1963
- (b) 1968   *
- (c) 1969
- (d) 1974

8 The Earth's nearest planetary neighbour is:
- (a) the Moon
- (b) Mercury
- (c) Mars
- (d) Venus   *

9 Mercury and Venus are not suitable for life as we know it because:
- (a) they are too cold
- (b) they are too hot

- (c) of lethal ultraviolet radiation
- (d) (b) and (c)   *

10 Modern humans have inhabited the planet for about:
- (a) 45 million years
- (b) 4.5 million years   *
- (c) 45 thousand years
- (d) 4.5 thousand years

11 In the seventeenth century the age of the Earth was accepted as:
- (a) 300 thousand years old
- (b) 6 thousand years old   *
- (c) 6 million years old
- (d) 300 million years old

12 James Lovelock developed a computer model called:
- (a) Daisyworld   *
- (b) Gaia
- (c) Geophysiology
- (d) Biosphere

Figure questions

1   Figure 3.8 illustrates the revolution of the Earth around the Sun. Answer the following questions.

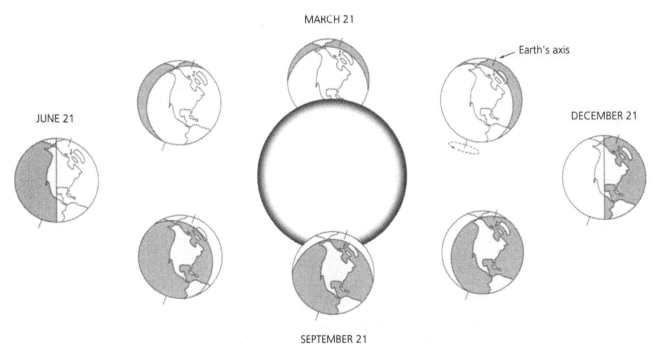

Figure 3.8 Revolution of the Earth around the Sun. The Earth revolves around the Sun over a yearly cycle, and this defines the passage of the seasons in both hemispheres. See text for explanation. After Figure 2.2 in Doerr, A.H. (1990) Fundamentals of physical geography. *Wm. C. Brown Publishers, Dubuque*

(a) What cycles illustrated in the diagram affect climate?

(b) How do these cycles determine the amount of solar energy received at the Earth's surface?

(c) What other cycles affect a place in relation to the Earth's position relative to the Sun?

Answers

(a) The yearly orbit of the Earth around the Sun determines seasonal climate. The daily rotation of the Earth about its axis will provide diurnal change, in local climatic conditions, provided a place passes through the plane of day–night during this daily cycle.

(b) The height of the Sun in the sky is determined by the yearly cycle producing a seasonal rhythm. This affects the angle at which the Sun's rays intersect the ground and thus the area of surface heated per unit of solar energy. The length of day is determined by the tilt of the Earth's axis relative to the Sun. Thus in December the south polar area has a 24-hour day. In this instance the duration of solar energy determines the amount received. Finally, the tilt of the Earth relative to the Sun determines the depth of atmosphere the solar energy has to pass through.

(c) The Earth's precession (tilt of the Earth relative to the plane on which the Earth moves around the Sun). The obliquity of the Earth (tilt of the Earth's axis of rotation). Both these have cycles of tens of thousands of years. At the timescale of hundreds of thousands of years the eccentricity of the Earth's orbit around the Sun has a cycle changing from more circular to more elliptical and back.

Short-answer questions

1 Briefly describe the Earth as a system.

Answer

The Earth operates as a closed system, except for some minimal exchanges of material between the upper atmosphere and space. The Earth has inputs and outputs of energy but not materials. Energy is received from the Sun as mainly short-wave radiation and outputted back to space mainly as long-wave radiation.

2 How did Pythagoras view the Earth?

Answer

Pythagoras viewed the Earth as a sphere but he saw it as stationary within space. He had an Earth-centric view with the Earth positioned at the centre of the Universe. The Earth itself was seen as dynamic being subject to change but essentially this was only of decay through time.

3 How did Copernicus view the Universe?

Answer

Copernicus positioned the Sun at the centre of the Universe. He identified the Moon as an orbiting satellite of the Earth, with the Earth and other planets following a concentric circular path around the Sun.

4 Outline the links between sunspot cycles and environmental changes on Earth.

Answer

Whilst having a great deal of short-term variation, over the longer term sunspots exhibit a semi-regular 11-year cycle. This cycle has been correlated to long-term rainfall records, with maximum rainfall occurring at times of greatest sunspot activity. There is evidence over much longer time-scales that they may influence long-term climate change. Quickening of the shorter-term cycle, effectively increasing sunspot activity per decade, has a positive association with increased global temperatures. Some argue that sunspot activity may, therefore, have more influence on recent climate change than the enhanced greenhouse effect.

5 Outline the Big Bang theory.

Answer

The Universe evolved from a super-dense concentration of matter that suffered a cataclysmic explosion. The observed expansion of the Universe is a result of this explosion.

6 What techniques may be used to date rocks over the geological timescale?

Answer

Relative dating techniques of rocks may be determined by evidence of fossils within them (palaeontology) and by their positions in a geological sequence (stratigraphy). However, the measurement of the decay of radioactive substances in rocks allows for absolute dating. Radiocarbon dating can date sediments back about 30,000 years, whilst potassium-argon dating can cover up to 30 million years.

7 Describe the Earth's time zone system.

Answer

The globe is divided into 24 one-hour time zones. Each one covers 15° of longitude. The time zones are relative to the zero meridian at Greenwich (Greenwich Mean Time). Zones to the east are thus sequentially one hour in advance and those to the west similarly one hour behind.

8 What is an equinox?

Answer

This is when the Sun appears directly overhead at midday on the equator. This occurs twice yearly on 21 March (the vernal equinox) and 23 September (the autumnal equinox).

9 Where does the Gaia hypothesis position human society?

Answer

The Gaia hypothesis offers a global and evolutionary view which puts human society into its context as part of the natural Earth system. Humans are displaced from the centre of the Earth's purpose and function. This view demands respect for other species and the natural systems that support life.

10 What is geophysiology?

Answer

It is a term used by James Lovelock to describe evolution of organisms and their material environment as a symbiotic single process.

Additional references

Kargel, J. S. and Strom, R. G. (1996) 'Global climatic change on Mars'. *Scientific American* 275(5), 60–8.
A speculative account of past climatic change on Mars and its implications for surface processes as well as inputs from outside the planet's system. Well illustrated, indicating system interactions and outcomes.

Web site

www.nosc.mil/planet-earth/environment
The home page of 'Planet Earth' that provides dynamic links to a wide variety of global sites. Useful links to initiatives based on the Gaia hypothesis.

Aims

- To reinforce the concept of long-term and large-scale change as part of the natural order of things.

- To examine the structure of the Earth's interior and its relationship to the surface.

- To illustrate the interior–surface relationship with evidence from environmental problems (radon and nuclear waste) and applications (geysers and geothermal energy).

Key-point summary

- The balance of land and water on the Earth's surface is unequal in both size and distribution. This distribution is constantly dynamic due to long-term processes for change. The major processes for change relate to *plate tectonics* and associated activity such as *earthquakes* and *volcanoes*.

- The spatial extent of the oceans will also be affected by *sea-level change* responding to *climate change*.

- The differential heating and cooling of the land masses and oceans has a direct influence on global patterns of climate.

- Human systems are also affected by and impinge on this distribution of land and water. The global distribution of pollution is a good example of this interrelationship.

- Comparison of the depth of the Earth's interior divisions in relation to the surface relief clearly demonstrates the restricted confines of the human living space. Further scale analogies between people and these relief features reinforce the view of the human species as a small evolutionary component of the Earth.

- The internal structure of the Earth is far from uniform, comprising a series of layers (zones) built up on each other. These layers affect the exterior surface of the Earth, via an interacting system. Though our picture of the interior delineation of the Earth is drawn from indirect evidence, there is a general consensus regarding the properties of the various zones. These are presented as:

1 the core, containing mainly iron and nickel, and outer zones. A solid *inner core* at extremely high temperatures (4,500–5,500°C), and a surrounding molten outer core cooling to less than 2,000°C at its outer boundary. The key influence of the core, felt at the surface, is the generation of the Earth's magnetic field by the rotation of the molten outer core about the solid inner core;

2 the *mantle*, again comprising two zones. The *asthenosphere* consisting of semi-fluid rocks. Convection currents within the asthenosphere are the driving force behind the redistribution of land and sea via the mechanism of plate tectonics. The second and outer zone is the *lithosphere*. This rigid material constitutes a layer of strength relative to the deformable asthenosphere below;

3 the *crust* which is the outer skin of the Earth. It merges with the lithosphere and the boundary between the two is of variable depth. Some literature amalgamates the two terms as the lithosphere.

- The crust is comprised of both oceanic (*sima*) and continental (*sial*) crust. They have different thicknesses, densities and weights. This affects their relationship to each other and is interdependent with plate tectonic processes.

- *Radioactivity* is naturally present in earth materials. An important source affecting humans is *radon gas*. It is released by weathering or tectonic processes and links to the groundwater component of the hydrosphere as it is dissolvable. Its occurrence is spatially variable, dependent on source and system links to the surface via water and fissuring. Radon is a health risk when it accumulates in confined spaces usually related to human resource systems, e.g. mines and housing.

- Nuclear waste storage underground may be considered as the waste flows of material to the geological store. In effect it is the opposite of the natural system flow. Lack of knowledge and foresight has produced the present problem. The spatial distribution of nuclear waste is uneven due to discrete development of nuclear power, and this distribution does not necessarily correspond to the suitable sites for underground storage. It must, therefore, be transported (flow) through the biosphere for safer disposal.

- Hot springs and geysers link underground processes and water stores to the surface. They display a frequency of eruption that indicates patterns of process in other environmental systems (seismic, atmospheric and tidal). This produces an irregular eruption pattern contradicting popular held beliefs (folklore).

- Below surface, conditions may be useful as well as destructive. *Geothermal energy* may be exploited at selective points on the surface corresponding to near-surface (about 10 km depth) molten rocks conducting heat upwards. As with other renewable energy sources environmental impacts relate to the unsympathetic development of the resource.

Main learning hurdles

The Earth's outer layers

At this stage terminology and spatial relations between the components of the outer layers of the Earth's structure must be clarified. Otherwise, the principles of Plate Tectonic Theory (Chapter 5) may be difficult to grasp. Crustal type (sima and sial) differences and their characteristics must be emphasized, particularly in relation to isostatic balance. The crust must be emphasized as being coupled to the upper mantle and that these two comprise the lithosphere. They are uncoupled from the asthenosphere by the non-uniform depth of partially molten mantle below. This allows for associated plate movement. The Moho is a boundary within the lithosphere between the crust and mantle.

Seismic waves

A reasonably comprehensive knowledge of seismic waves is important as evidence concerning our knowledge of the Earth's interior and also in relation to the variable effects each type of wave has on the surface system. Both speed and wave movement need emphasis. A good way of demonstrating the types of wave is to use a stretched 'slinky' spring.

Key terms

Core; crust; earthquake prediction; geodesy; geothermal gradient and energy; geysers; lateral recharge; mean sea-level; mantle; mountain building; nuclear waste; plate tectonics; radioactivity; relief; viscous; seismic waves; Sellafield; sial; sima; sustainable energy; radon; Yucca Mountain.

Issues for group discussion

Discuss the safety of underground nuclear storage

The discussion should focus on the system variables involved, e.g. rock type, strength and integrity. Spatial pattern factors should be drawn out in relation to geological stability (earthquake zones) and human settlement. These dissimilar distributions should allow the discussion to evaluate the safety of nuclear waste transportation.

Discuss the views that geothermal energy is environmentally friendly

Students should at least read Scudder (1990). The teaching emphasis should be on developing an holistic argument, relative to other sources of energy, based on scale of impact at local, regional and global levels. Students should prepare principally an environmental argument but recognition of the economic costs need monitoring regarding possible implementation.

Selected reading

Jeanloz, R. and Lay, T. (1993) 'The core-mantle boundary'. *Scientific American* 268(5), 26–33.
An article emphasizing the importance of interfaces between layers within the Earth. They affect the Earth's rotation and magnetic field with influences

on other systems. A good synthesis of material from theories and scientific evidence is presented to illustrate the composition and inner workings of the Earth.

Scudder, B. (1990) 'Energy galore'. *Geographical Magazine* 62(9), 40–4.
Introductory account of the technology behind Iceland's geothermal energy exploitation. Clearly demonstrates the economic as well as the environmental argument.

Silver, P. G. and Valetto-Silver, N. J. (1992) 'Detection of hydrothermal precursors to large northern California earthquakes'. *Science* 257(5057), 1363–8.
Spatially equates geothermal surface evidence with underground seismic activity. Applications relating to earthquake prediction are discussed in relation to other physical factors. Useful graphical illustration.

Textbooks

Berkhout, F. (1991) *Radioactive Waste: Politics and Technology*. Routledge: London.
Compares management of radioactive waste using case studies from Europe. A clear focus presents the technological issues in a policy framework.

Clark, M., Smith, D. and Blowers, A. (eds) (1992) *Waste Location: Spatial Aspects of Waste Management, Hazards and Disposal*. Routledge: London.
A series of topical essays on waste management from a spatial perspective. Nuclear waste is considered in a number of thematic areas with a significant chapter on the criteria needed for effective management of radioactive waste.

Goudie, A. (1993) *The Nature of the Environment*, 3rd edn. Blackwell: Oxford.
A clear, student-friendly introductory text for undergraduates. Major environmental systems are set in a global framework comprising both a geological and climatic background. A wide variety of landscapes and ecosystems is evaluated in the context of major dynamic processes and natural physical cycles. Well illustrated with key ideas clearly flagged.

Hill, R., O'Keefe, P. and Snape, C. (1995) *The Future of Energy Use*. Earthscan: London.
Explores in a clear and thorough manner the technical and economic issues of a wide variety of energy sources covering conventional, alternative and nuclear power. Sustainable futures applicable to both national and planetary systems are presented with an interdisciplinary environmental focus. This provides an assessment of the difficulties involved in providing an integrated costing of the effects on global environmental systems.

Openshaw, S., Carver, S. and Fernie, J. (1989) *Britain's Nuclear Waste: Safety and Siting*. Belhaven: London.
A book that raises many issues. The evidence is presented in an objective fashion allowing readers to formulate their own views. There is comprehensive coverage of the scientific aspects of the nuclear waste issue as well as the politico-economic dimension. This is presented in an integrating spatial analysis.

Park, C. C. (1989) *Chernobyl: The Long Shadow*. Routledge: London.
This temporal study of the Chernobyl incident and its future implications are discussed from an interdisciplinary perspective. The issues are explored in a highly readable manner. Scientific data, public perceptions and political reactions are major themes drawn together to contextualize the contribution of this major environmental issue. Students are able to consider the interconnections between 'hard science' and social policy.

Summerfield, M. A. (1991) *Global Geomorphology*. Longman: Harlow.
A comprehensive introduction to large-scale processes and landforms. There is a clear integration of global plate tectonics with major landform development. This is underpinned by a review of the relationships between the internal and external processes relevant to the development of the Earth's surface.

Essay questions

1 Discuss the main characteristics of the Earth's crust.
2 Outline the evidence used for determining the internal structure of the Earth.
3 Explain the principle of isostasy.
4 Discuss the distribution of the radon hazard.
5 'Natural radiation is a greater hazard than radiation from human sources.' Evaluate.
6 What are the problems associated with storing nuclear waste underground?
7 Outline the natural mechanisms generating

geothermal heat and explain its potential for exploitation spatially.

8   Evaluate the use of geothermal energy as an energy source.

Multiple-choice questions

Choose the best answer for each of the following questions:

1   The branch of science that measures the size, shape and curvature of the Earth is:
   (a) geodesy
   (b) cartography   *
   (c) geophysics
   (d) geomorphology

2   How much of the Earth's surface is occupied by land?
   (a) 9 per cent
   (b) 29 per cent   *
   (c) 49 per cent
   (d) 69 per cent

3   The highest peak in Europe is:
   (a) the Matterhorn
   (b) the Eiger
   (c) Mont Blanc
   (d) Elbrus   *

4   The deepest wells extend down:
   (a) 0.1 km
   (b) 1 km
   (c) 11 km   *
   (d) 110 km

5   Which of the following contributes most to our knowledge of the Earth's interior:
   (a) sedimentology
   (b) remote sensing
   (c) seismic waves   *
   (d) mountain building

6   The crust is approximately:
   (a) 27 km thick   *
   (b) 60 km thick
   (c) 290 km thick
   (d) 2,900 km thick

7   The Earth's inner core has an estimated temperature of:
   (a) 450–550 degrees centigrade
   (b) 4,500–5,500 degrees centigrade   *

   (c) 1,500–2,500 degrees centigrade
   (d) 11,500–12,500 degrees centigrade

8   Ocean crust is:
   (a) the same age as continental crust
   (b) older than continental crust
   (c) of a similar age to the Moon
   (d) younger than continental crust   *

9   The oldest parts of the continental crust are:
   (a) shield areas   *
   (b) the Andes
   (c) sedimentary basins
   (d) diamond mining areas

10   Which of the following is the most common element in the Earth's crust:
   (a) iron
   (b) silicon
   (c) oxygen   *
   (d) nitrogen

11   Earth materials account for how much natural background radiation:
   (a) 60 per cent   *
   (b) 50 per cent
   (c) 40 per cent
   (d) 30 per cent

12   Radon is produced by the natural decay of:
   (a) uranium
   (b) plutonium
   (c) radium   *
   (d) helium

13   Which of these is not a characteristic of alpha particles:
   (a) a positive charge
   (b) high ionization properties
   (c) high penetration rate   *
   (d) high rate of energy loss

14   High radon concentrations are associated with:
   (a) limestone
   (b) chalk
   (c) gneiss
   (d) granite   *

15   High-level nuclear waste needs safe storage for up to:
   (a) 500 years
   (b) 5000 years
   (c) 10,000 years   *
   (d) 100,000 years

16 The Yucca Mountain is in:
(a) Oregon
(b) Virginia
(c) Nevada   *
(d) Arizona

17 Temperature rise per kilometre of depth below the Earth's surface, known as the geothermal gradient, is on average:
(a) 2–4 degrees centigrade
(b) 20–40 degrees centigrade   *
(c) 12–14 degrees centigrade
(d) 5–10 degrees centigrade

18 By the early 1990s geothermal energy was being used in:
(a) 2 countries
(b) 4 countries
(c) 8 countries
(d) 20 countries   *

19 To produce 500 megawatts of electricity from geothermal energy on Hawaii would need a scheme costing:
(a) US $400,000
(b) US $4,000,000
(c) US $400,000,000
(d) US $400,000,000,000   *

20 Geothermal energy exploitation from active volcanic zones may release:
(a) hydrogen sulphide   *
(b) nitrogen dioxide
(c) carbon dioxide
(d) argon

Figure Questions

1 Figure 4.5 shows a section through the Earth. Answer the following questions.
(a) What are the constituents of the indicated layers of the Earth's interior.
(b) On the diagram indicate the movement of seismic waves from the earthquake focus.
(c) Outline the characteristics of the two main types of seismic waves.

Answers

(a) The inner core is solid and is made up of mainly iron with some nickel. Silicon and sulphur may be present. The outer liquid core is comprised mostly

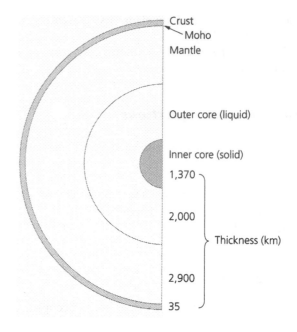

Figure 4.5 *The Earth's mantle and crust. Beyond the thin crust is the relatively thick mantle. The Earth's core is solid, with a liquid outer part. After Figure 1.1a in Goudie, A. (1993) The nature of the environment.* Blackwell, Oxford.

of molten iron and nickel. Composition of the mantle varies with distance from the core. It is comprised of minerals (silicate compounds) and metals (iron and magnesium). Near the core they are in a semi-fluid state, whereas in the upper part of the mantle they are rigid.

(b) Refer to Figure 4.4.

(c) The two main types of seismic waves are p-waves and s-waves. P-waves are able to travel through both solid and liquid, being refracted as they pass through the liquid outer core. S-waves are only able to pass through solids.

2 Figure 4.8 shows the entry points for radon gas into a home. Answer the following questions.
(a) Describe the system pathways into the house.
(b) Where is radon likely to concentrate?
(c) What are the health risks to humans posed by radon gas?

Answers

(a) Radon enters the home as a gas from weathered and cracked rocks. It passes into the house via physical gaps in the foundations adjacent to the

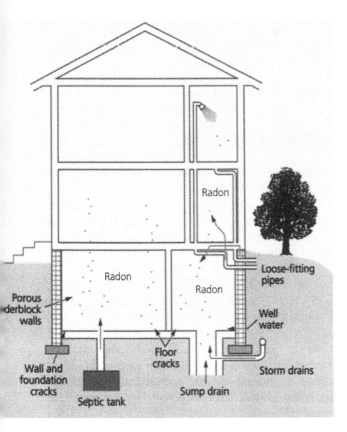

Figure 4.8 *Main entry points for radon gas into a house. Radon gas, which is naturally present in some rocks, can enter buildings in a variety of ways. Most involves seepage of the gas through cracks, porous material, pipes and drains. After Figure 9.12 in Marsh, W.M. and J.M. Grossa (1996)* Environmental geography: science, land use and Earth systems. *John Wiley & Sons, New York*

ground surface. Radon gas may enter other rooms by gaps in the floorboards. When dissolved in water it may enter directly by passage of soil water through porous lower walls or indirectly via the water supply system. The internal domestic water system may then distribute it around the house to escape via taps and shower heads.

(b) Radon is likely to concentrate in the basement as a combination of gaseous and soil water entry. There will also be a lower concentration in bathrooms and kitchens.

(c) Radon has been suggested as a causal factor of lung cancer and leukaemia.

Short-answer questions

1   Describe the Earth's crust.

Answer

The crust is the outer veneer of the Earth. It varies in thickness between 8 km and 65 km. It comprises oceanic crust made up of silicon and magnesium (sima) and continental crust made up of silicon and aluminium (sial).

2   What are the extremes of the Earth's relief?

Answer

On the surface of the Earth the highest point is the summit of Mount Everest, in the Himalayas. It is 8,872 m above sea level. The lowest point is the Mariana Trench, in the Pacific Ocean, at a depth of 11,034 m. The total relief system is therefore only about 20 km in range or about 0.3 per cent of the Earth's radius.

3   What is the Moho?

Answer

Moho is short for Mohorovicic discontinuity. It is a sharp, clearly defined boundary between the crust and mantle. Therefore, its depth varies with the thickness of crust, being about 32 km below the continents and 10 km below the ocean floor.

4   Highlight the differences between oceanic and continental crust.

Answer

Oceanic crust is thinner and relatively denser and heavier than continental crust. It consists mainly of basaltic rocks whereas continental crust is composed of granitic rocks. Oceanic and continental crust both contain silica but the other main chemical constituent differs: magnesium in oceanic crust is replaced by aluminium in continental crust.

5   Suggest two ways in which the raw materials of the Earth's crust link to other systems.

Answer

Firstly, the Earth's crust provides basic resources to the human economic system. Rocks provide building materials, fossil fuels provide energy resources, and metal and minerals are used in manufacturing. Secondly, they interact near the surface with weathering processes to play important roles in the biogeochemical cycles important to the workings of the biosphere.

6   What is a radioactive isotope?

Answer

An isotope is one of a number of forms of the same element whose atoms contain the same number of electrons and protons but different numbers of neutrons. Therefore, they share common chemical properties but have differing physical ones. Radioactive isotopes are unstable. Stability is achieved by the emission of radioactive particles by radioactive decay, whose rate determines the half-life of the isotope.

7   Describe the ways heat is generated within the Earth.

Answer

Heat may be generated as a by-product of the radioactive decay of rocks. This mainly occurs at depths between 40 km and 48 km. The tremendous compression of material at great depths generates heat. Heat was generated during the initial formation of the Earth and some still remains in the interior as residual heat.

8   List the ways, and where, geothermal energy is being exploited.

Answer

Geothermal energy is used for direct local domestic heating in Iceland, the USA and Russia. It is used by industry in New Zealand, Russia and China. It provides heating in the food production sector (agriculture, greenhouse cultivation, aquaculture) in Hungary, China and Russia. It is also used for bathing in Iceland and Japan.

Additional references

Edwards, R. (1996) 'Sellafield's Trojan horse'. *New Scientist* 19 (2011), 11–12.
Explores the arguments against underground nuclear waste storage in Cumbria from the structural geological angle. Evidence built up from seismic surveys is discussed.

Van der Hilst, R. D., Widiyantoro, S. and Engdahl, E. R. (1997) 'Evidence for deep mantle circulation for global tomography'. *Nature* 386 (6625), 578–84.
Evidence from seismic wave travel times and the location of earthquakes suggests convective flows within the mantle. Spatial correlation of structures within the mantle and plate convergence zones are discussed.

Zorpette, G. (1996) 'Hanford's nuclear wasteland'. *Scientific American* 274 (5), 72–6
Considers the problems associated with nuclear waste storage. Clearly emphasizes the limitations of scientific knowledge, present technology and political agreement in dealing with the problem. Proposes that a time delay in appropriate action is storing up potential economic and environmental disaster for future generations.

Web site

www.geo.ucalgary.ca/vl-earthsciences
This site provides a comprehensive set of resources and some usable software. There is a range of dynamic links related to geology and its applications.

CHAPTER 5
Dynamic Earth

Aims

- To build on the concepts of the last chapter relating to the evidence we have about how the Earth works. Relating the internal evidence with the surface effects.

- To examine time-scales of change. In particular to differentiate between the long-term global processes, of plate tectonics and mountain building, with the short-term changes related to geological hazards (volcanoes and earthquakes).

- To emphasize that these processes have both a physical topographic outcome and a human system outcome.

Key-point summary

- The global landmass has split up and produced the continents which have moved through time. A variety of evidence supports this view with associated refining of theory:

  1 observational evidence of opposing continental outlines, especially South America and Africa, shows a strong visual shape correlation;

  2 spatial positional evidence from *Continental Drift Theory*, including the location of ancient deposits indicating specific climatic conditions no longer associated with their current global position. Fossil evidence from spatial disparate land masses adds further support;

  3 palaeomagnetic evidence suggesting that continents have moved on the Earth's surface relative to each other in the past.

- A mechanism for the movement is suggested by *Plate Tectonic Theory*. Wegener's proposition of floating continental crust (*sial*) on a plastic lower layer of denser crustal rocks (*sima*) had no ordered powering system initiating the break up of the super-continent. *Sea-floor spreading*, the lateral expansion of the ocean floor from the *mid-ocean ridges*, drives the movements of the continental crust over the ocean crust.

- This crustal plate movement shapes the large-scale relief features on the surface of the Earth. The dynamics of this generally occur in the vicinity of *plate boundaries*. Here different types of plate interaction set up forces that produce different outcomes of both relief via mountain building, and environmental hazard, via seismic activity. Plate tectonics is seen as important by providing a unifying theory that links these systems, at the large scale, in a spatially related way.

- At a scale below this basic framework discrete relief features, e.g. mountains and hazard events, e.g. volcanic eruptions and earthquakes, interact with the human system enabling or disrupting human communities and activities.

- Subsequent deformation of the crust may also be viewed at a variety of spatial scales. *Warps*, large regional-scale bends of the Earth's crust, have smaller-scale *folds* and *faults* superimposed upon them. The directionality of the forces producing these forms also serves to differentiate between them. Here generalizations from plate tectonic theory provide a link between form and process. Warping is a response to vertical pressure such as the weight of sediment accumulation and folding is a response to horizontal movement e.g. plate convergence. Faults are related to horizontal and vertical displacement of the rocks adjacent to plate boundaries. Here pressure (compressional or tensional) exceeds the plastic limit (*threshold*) of the rock and fracture occurs.

- Resultant landforms reflect this hierarchical scale

of deformation. Warping produces the largest area features such as *synclinal* ocean basins. Followed in scale by folding which produces mountain chains and finally fault systems which produce *graben* and *horst*.

- Plate tectonics and associated fault line movement explain the causes and mechanisms of earthquakes. Restricted movement at fault boundaries stores strain energy. Rocks adapt to this by bending until a threshold is reached and there is a sudden failure. The more accumulated energy, a product of time, the larger the earthquake.

- Evidence from monitoring and records suggests earthquakes are not random in timing and that these hazard events have a specific cycle of activity.

- The general patterns of major earthquakes and volcanoes equates to the tectonic plate boundaries. Increased human settlement in these areas has increased the number of disasters related to these hazards.

- Damage caused by earthquakes does not necessarily equate to the energy released by the earthquake. Smaller aftershocks of greater frequency and time duration may be more damaging. They also compound the damage produced by the initial shock. This sequence may be further extended by the triggering of other geological hazards such as landslides. The earthquake hazard may be viewed, therefore, as an interlinking series of events connecting geological, topographical and human systems. Most damage largely depends on the human system attributes affected by the force of the shock. The main attributes being the size of population adjacent to the *epicentre*, the density of infrastructure and the quality of this infrastructure. Consequently, there is not always a loose correlation between the output (earthquake energy) from the geological system and the magnitude of human disaster.

- Human settlement in earthquake zones necessitates the need for reducing the effects of the hazard. This may be approached in a number of ways, with varying degrees of success:

1   engineering solutions to produce earthquake-proof structures. A direct response to the earthquake energy that is effective but reliant on money available to implement successfully;

2   land use mapping to identify the areas most at risk. This may be a direct spatial response to the likely location of earthquakes. More importantly it relates to the interaction of the shocks with other surface systems such as soil and slope (secondary hazards). Zones at risk may be identified and used for low-density human activity. Good scientific knowledge of surface process reaction to shock makes this effective if still ultimately dependent on human choice of residential location;

3   increased monitoring of earthquake zones in an effort to predict the location, timing and size of an earthquake event. This information allows for emergency measures to be implemented. Limited long-term accuracy of the timing and size of individual events make its usefulness at present problematic.

- Human interference in the earthquake zone may increase the likelihood of earthquake hazards by triggering fault line failure due to crustal loading (*reservoirs*), lubrication of inactive faults (sub-surface liquid storage) and underground nuclear testing.

- An understanding of the dynamism at plate boundaries displayed by the internal systems within the Earth aid our knowledge of volcanic activity. Spatial distribution, like earthquakes, generally corresponds to plate margins. Although spatial anomalies occur they are difficult to account for.

- In contrast to the suggested periodicity of earthquake activity volcanoes display a random behaviour through time.

- Like earthquakes, volcanic eruptions and associated hazards appear to have increased in number during the twentieth century.

- Our evidence to produce an understanding of the processes of vulcanism is essentially confined to the surface effects of the internal mechanisms.

- *Magma* and *lava* are the two basic volcanic materials. Magma is produced by the partial melting of rock in the asthenosphere. This may be due to a decrease in pressure and/or an increase in temperature. Lava is magma that flows out of vents and fissures on the surface system over the short time-scale by exuding molten material on to the landscape system and releasing gases into the atmospheric system. Over longer time-scales the cooling of the material produces hard igneous rocks influencing the surface relief system. The nature of the volcanic material influences its relationship with surface systems. Chemical composition (particularly *silica*) controls the resistance to flow (*viscosity*) and thus the spatial coverage of the rock on the surface. It also controls the resistance of the rock to global weathering systems interrelating with relief and soil systems.

- *Vulcanism* takes place within the internal system below the Earth's surface (*intrusive*) and via pathways on the Earth's surface (*extrusive*). A variety of identifiable discrete forms result. Dynamic surface weathering processes link the two by eventually uncovering intrusions so that they become surface features.

- Fissuring and fractures in the overlying heavier rocks allow the lighter magma the potential for access to the surface.

- Volcanoes are the surface manifestation of this rising magma. Their character depends on both mode of formation and the characteristics of the material involved (lava).

- Volcanic hazards vary in their spatial impact and the directness of their action. Primary hazards such as *ash fall-out* and *lava flows* tend to be local in extent. Local and regional seismic disturbance may produce secondary hazards such as *landslides* and *mudflows* within other surface systems. On a larger scale material may be inputted into the atmosphere where long-lasting climatic effects result.

- Evidence from previous volcanic events allows the spatial mapping of hazard zones. Application of this knowledge produces appropriate land-use planning to reduce the hazard risk and provide more effective disaster response. Prediction of impending volcanic eruption is more reliable than earthquake prediction as volcanic vents are spatially discrete. However, at the local level unpredicted directional responses may occur such as the Mount St Helens lateral blast.

Main learning hurdles

Plate tectonic theory

Basic principles related to the near surface structure of the Earth must be reviewed from the previous chapter. The spatial relationship of features requires emphasis and a reference to the Earth's surface topography without oceanic waters should clarify the distribution of processes associated with plate tectonics. Explanation of the application of plate tectonics to provide reasons for large-scale geographical form (continental shape) and processes (earthquake zones) should be attained initially. From this the concept of time-scales may be introduced related to contemporary change at plate margins and longer timescale implications of the build up of continental crust. For some students used to true north on maps and the idea of fixed magnetic polarity the palaeomagnetic evidence for plate tectonic process and continental movement should be examined thoroughly.

Isostatic and eustatic adjustment

The relative vertical movements of continental crust and its relation to sea-level change can be a difficult concept for students to grasp thoroughly. As well as the directions of movement the rate of movement must also be explained so that the student can understand that sea-level change may be experienced given uni-directional movement of both land and sea. The students must be clear about the models of isostasy and eustasy.

Earthquake measures

The instructor must clearly differentiate between the rationale of the Richter and Mercalli scales, emphasizing the physical premise of Richter and human-effect focus of Mercalli.

Key terms

Anticline; constructive and destructive boundaries; continental drift; cratons; diastrophism; dome; earthquake focus; elastic strain; epicentre; extrusive and intrusive vulcanism; faulting; folding; fracturing; graben; horst; isostatic adjustment; landslides and mudslides; Laurasia; lava; magma; magnitude; Mercalli; mid-ocean ridges; monocline; mountains; orogenesis; palaeomagnetism; plastic strain; plates; plate tectonics; plutons; polar wandering; reverse fault; Richter; sills; subduction; synclines; thrust fault; transcurrent fault; tsunami; viscosity; volcanoes; warping; Wegener.

Issues for group discussion

Discuss the global distribution of mountains

The students should be encouraged to focus on a number of examples and scales of mountain ranges.

Introductory reading from Harrison *et al.* (1992) should be used to link mountain buildings with tectonic processes. The discussion should develop a chronology of processes and relate the case studies of mountain chain relief to this time sequence.

Discuss the factors affecting the human loss experienced from a major earthquake

Following a spatial introduction such as Tyler (1990) a variety of case studies may be used to develop the discussion. The instructor should ensure that examples from both the economically developed and economically underdeveloped worlds are covered.

Selected reading

Dorak, J. J., Johnson, C. and Tilling, R. I. (1992) 'Dynamics of Kilavea Volcano'. *Scientific American* 267 (2), 18–25.
Comprehensive scientific monitoring of Kilavea is the basis of understanding volcanic workings and subsequent predictions of surface effects. Well illustrated local/regional spatial distributions of system pressures are explored.

Freet, S. (1992) 'The deadly cloud hanging over Cameroon'. *New Scientist* 135 (1834), 23–7.
An account of the hazard outcome of natural toxic gas release. The linking of internal systems with both hydrological and climate systems usefully suggests the causes and explains the spatial influence of the hazard.

Fumal, T. E., Pezzopane, S. K., Weldon, R. J. and Schwarz, D. P. (1993) 'A 100-year average recurrence interval for the San Andreas Fault at Wrightwood, California'. *Science* 259 (5092), 199–203.
Uses sedimentological evidence of previous earthquakes to estimate recurrence intervals. The evidence is presented in clear sectional diagrams. Risk estimates and predictions are discussed from the presented observational evidence.

Harrison, T. M., Copeland, P., Kidd, W. S. F. and Yin, A. (1992) 'Raising Tibet'. *Science* 255 (5052), 1663–70.
Multidisciplinary studies and evidence are used to explain tectonic evolution of the Tibetan region. A chronology is built up and a general model proposed.

Kirkbride, M. P. and Sugden, D. E. (1992) 'New Zealand loses its top'. *Geographical Magazine* 64 (7), 30–4.
A case study of rock avalanches in New Zealand. Emphasizes the unpredictability of these events and usefully illustrates the mechanics behind them. Excellent links to their effects on glacial system mass balance.

Oppenheimer, C. (1990). 'Monitoring hot spots from space'. *Geographical Magazine* 62 (2), 32–4.
Clearly emphasizes the scale of volcanic activity and explains why comprehensive coverage of these events requires remote sensing technology.

Romanowicz, B. (1993) 'Spatiotemporal patterns in the energy release of great earthquakes'. *Science* 260 (5116), 1923–6.
A concise classification of major earthquake events by fault mechanism. Global patterns linking plate tectonics and large earthquake events this century are displayed in map form.

Roy, A (1990) 'Countdown to the big one'. *Geographical Magazine* 62 (12), 44–8.
Case study article of the extensive earthquake monitoring system at Parkfield, California. Suggests reasoning behind predictive studies. This article can be usefully related to Californian earthquakes of the 1990s.

Tyler, C. (1990) 'Earth-moving events'. *Geographical Magazine* 62 (3), 28–31.
Focuses on earthquake disaster preparation. Well illustrated spatial relationship between major urban centres and plate margins.

Textbooks

Alexander, D. (1993) *Natural Disasters*. UCL Press: London.
Very useful section on geographical agents. Strong on discussing intensity of hazard effects, monitoring systems and predictive capabilities.

Chester, D. K. (1993) *Volcanoes and Society*. Arnold: London.
A clear outline of the workings of volcanoes. Very useful sections on the human response and adaptation to the volcanic hazard environment.

Francis, P. (1993) *Volcanoes: A Planetary Perspective*. Oxford University Press: Oxford.

An informative text that provides a global overview of vulcanism. Dynamic inner Earth processes and their surface hazard manifestations are explained in a student-friendly, readable fashion.

McCall, G., Laming, D. and Scott, S. (1992) *Geohazards: Natural and Man-made*. Chapman and Hall: London.
A well-illustrated book that stresses the application of geomorphology in hazard appraisal. Excellent cases studies of volcanoes, earthquakes and secondary hazards such as landslides.

Ollier, C. (1988) *Volcanoes*, 2nd edn. Blackwell: Oxford.
Accessible account of vulcanism and the variety of associated landforms. Strong emphasis on correlating distributional patterns with other systems. Useful discussional case studies.

Smith, K. (1992) *Environmental Hazards*. Routledge: London.
Contains a comprehensive chapter on geological hazards. Clear links are explored in relation to physical systems and the human social system.

Yeats, R. S., Sieh, K. E. and Allen, C. R. (1997) *Geology of Earthquakes*. Oxford University Press: Oxford.
A comprehensively illustrated text drawing on well-documented case studies of twentieth-century earthquakes. Earthquake studies are approached from a multidisciplinary perspective.

Essay questions

1 Describe and account for the evolution of a fault scarp through time.
2 Earthquakes with high values on the Richter scale are not always the most devastating, nor do they cause the greatest loss of life. Why?
3 Describe and illustrate the processes occurring at different plate boundaries.
4 Explain the factors influencing sea level relative to the land.
5 Discuss how plate tectonic theory has helped explain processes at the global and regional scale.
6 Examine the concept of 'system equilibrium' in relation to oceanic and continental crust.
7 Evaluate the contention that the environmental impacts of short-term geomorphological processes are insignificant when compared with the processes which operate over longer time-scales.

8 Discuss the costs and benefits of living in an area of volcanic activity.
9 Describe and account for the main features associated with intrusive vulcanism.
10 How might we be able to predict future tectonic hazards?

Multiple-choice questions

Choose the best answer for each of the following questions.

1 Alfred Wegener put forward the theory of:
(a) plate tectonics
(b) palaeomagnetic anomaly
(c) continental drift *
(d) relativity

2 Continental shelves are part of:
(a) the ocean crust
(b) a mid-ocean ridge
(c) the continental crust *
(d) the asthenosphere

3 The theory of continental drift helps to explain:
(a) the fit of continental margins
(b) the location of old mountain belts
(c) shield area location
(d) (a), (b) and (c) *

4 India collided with Asia about:
(a) 5 million years ago
(b) 15 million years ago
(c) 50 million years ago *
(d) 350 million years ago

5 The plate tectonic model began to emerge in the:
(a) 1910s
(b) 1930s
(c) 1950s *
(d) 1980s

6 Plate tectonic theory was formalized in the:
(a) late 1920s
(b) late 1950s
(c) late 1960s *
(d) late 1980s

7 The average speed of plate movement is:
(a) 7 cm a week
(b) 7 cm a year *
(c) 70 cm a year
(d) 7 cm a decade

8 The Himalayan mountain chain is being raised by:

(a) collision and downwarp
(b) collision and upthrust   *
(c) divergence and upthrust
(d) divergence and downwarp

9 A large depression between two faults that are more or less parallel is a:

(a) graben   *
(b) syncline
(c) thrust plane
(d) basin

10 The upthrown block between two parallel faults is:

(a) an anticline
(b) a plateau
(c) a rift
(d) a horst   *

11 The Richter scale is a measure of:

(a) location of earthquakes
(b) magnitude of earthquakes   *
(c) intensity of earthquakes
(d) magnitude of volcanic explosion

12 An intensity measure of earthquake effects was devised by:

(a) Marconi
(b) Mercalli   *
(c) Wegener
(d) the Chinese

13 The easiest earthquake variable to predict is:

(a) timing
(b) size
(c) damage
(d) location   *

14 The distribution of volcanoes around the rim of the Pacific is known as the:

(a) Pacific Ring of Fire   *
(b) Pacific Rim of Fire
(c) Aleutian Islands
(d) Pacific Hot Spots

15 Lava that has solidified under water is known as:

(a) pillow lava   *
(b) aa
(c) pahoehoe
(d) basalt

16 Most explosive eruptions of volcanoes occur when the gas proportion of magma is at least:

(a) 25%
(b) 50%
(c) 65%
(d) 75%   *

17 The volcanic ash thrown out by an explosive volcanic eruption is:

(a) lapilli
(b) tephra   *
(c) pyroclastic
(d) cumulus

18 The Lake Nyas disaster was a result of:

(a) lava flows
(b) mudslides
(c) toxic gas   *
(d) volcanic ash

19 Most of the volcanoes in the Hawaiian islands are:

(a) ash core volcanoes
(b) solid core volcanoes
(c) plug dome volcanoes
(d) shield volcanoes   *

20 The most dangerous type of volcanic eruption is:

(a) Pelean   *
(b) Plinian
(c) Vesuvian
(d) Strombolian

Figure questions

1 Figure 5.7 shows the forms and processes at a typical subduction zone. Answer the following questions.

(a) What type of plate margin is illustrated and what is the major associated surface relief feature?
(b) Explain the composition of the major surface relief feature.
(c) What evidence might there be to indicate your answer to (b)?

Answers

(a) The continental-oceanic subduction zone is a destructive plate margin. The associated surface relief is a young fold mountain chain.

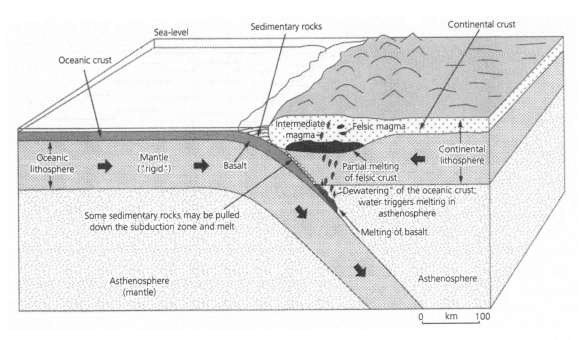

Figure 5.7 *Forms and processes at a typical subduction zone. One crustal plate is pushed downwards into the mantle beneath an adjacent one at a subduction zone, where some of the rock material is melted. After Figure 10.16 in Doerr, A.H. (1990)* Fundamentals of physical geography. *Wm.C. Brown Publishers, Dubuque*

(b) The young fold mountains are the result of the collision of the oceanic plate with the continental plate. As subduction takes place a trench is formed into which sediments collect and are compressed. These are uplifted to provide the ground mass of the mountains. The igneous volcanic zone may intrude and magma reaches the surface as volcanic extrusions. General system compression causes physical and chemical changes in areas of greatest pressure, resulting in metamorphic rocks.

(c) Evidence for sedimentary rocks may be marine fossils. There may be exposed granites or other plutonic rocks as igneous indicators. Active volcanoes indicate extrusive magma flow. There may also be folding evident, explaining the compressional forces.

2 Figure 5.18 shows the frequency of active volcanoes, 1860–1980. Answer the following questions.
  (a) Describe the trends shown on the graph.
  (b) Suggest reasons for these trends.
  (c) Outline the global distribution of volcanoes in relation to plate margins and the potential hazard posed by them.

Answers

(a) The general trend of the graph indicates an

increased number of active volcanoes between 1860 and 1980. Over shorter timescales, within this overall trend, there is a great deal of variability. However, a succession of peak values appears about every 30 years.

(b) The long-term increase may be due to an increase in natural volcanic activity. More likely it is a product of the increased coverage and sophistication of global monitoring and information dissemination. Short-term fluctuations probably reflect natural variability in activity. However, there are significant drops in the graph related to reduced scientific effort, resources and information transfer during the two world wars.

(c) Violent, explosive, hazardous volcanoes most frequently occur at convergent margins. Divergent margins produce less explosive shield volcanoes which are generally less of a hazard.

Short-answer questions

1 What disciplines utilize evidence from the theory of continental drift?

Answer

Palaeontology, the study of fossils, has used the

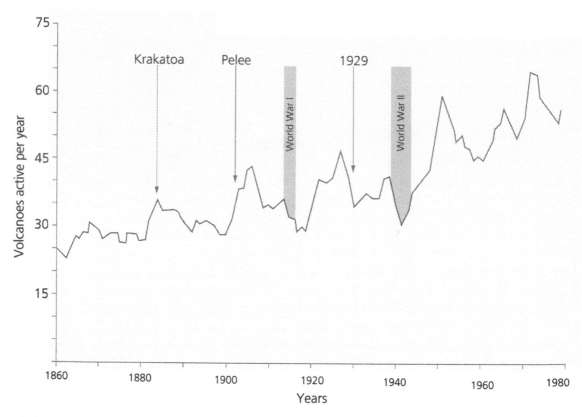

Figure 5.18   Increasing frequency of active volcanoes, 1860–1980. The graph, based on the three-year running average number of active volcanoes, shows a great deal of short-term variation superimposed on an underlying increase through time. After Figure 2.2 in Smith, K. (1992) Environmental hazards. *Routledge, London.*

break up of super-continental land masses to explain similar species separated by major oceans. Drifting continents explain fossilized tropical species found in present-day cold climates. Biogeographers use continental drift to explain similar distributions of species in spatially disparate parts of the globe with a range of climates. The morphology of continental coasts is explained by the fit between adjacent continents as emphasized by geomorphologists. They also explain relic landforms produced under different climates due to latitudinal drift of land masses. Geologists also explain similar geologies on opposite sides of wide stretches of ocean by super-continent break up.

2   Outline the differences between cratons and orogens.

Answer

Cratons, also known as shields, are old stable core areas of continental crust. Orogens comprise more mobile crust around the cratons. These are more affected by earth movements and intense heating exhibiting greater deformation.

3   Name and describe the movement at the types of boundary between crustal plates.

Answer

The three types of plate boundary are constructive, destructive and conservative. At the constructive boundary the two plates move apart (divergence). Plate movement towards each other (convergence) is experienced at destructive boundaries. At a conservative boundary plates slide past each other in opposite directions, described as a transform movement.

4   How does rock respond to pressure?

Answer

Rock may respond to pressure in one of three ways.

It may be deformed by applied pressure but returns to its original form when this pressure is released. This is the rock's elasticity. When a rock does not return to its original form but retains its deformation it is referred to as plastic deformation. If the applied pressure exceeds the plastic limits of the rock it fractures, losing its structural integrity.

5   Outline the Richter and Mercalli scales related to earthquakes.

Answer

The Richter scale measures the magnitude of an earthquake. It is a logarithmic scale describing the amplitude of a seismic wave, relating the amount of energy released. Each earthquake, therefore, has a fixed scale value. The Mercalli scale measures the intensity of an earthquake, its effect on people and structures. It is a numeric scale corresponding to given characteristics of effect. As the intensity of an earthquake differs with distance from its epicentre a single earthquake may be represented by different Mercalli numbers dependent on the effects experienced at different locations.

6   Describe the hazards associated with the 1906 San Francisco earthquake.

Answer

Initially violent Earth movements produced faulting affecting adjacent buildings. The faulting ruptured gas mains causing widespread fires in the wooden buildings. This caused the greatest loss of life. Subsidence badly damaged most of the remaining buildings as liquefaction of the underlying clay took place due to the initial shock and series of aftershocks.

7   How might earthquakes be controlled?

Answer

A number of strategies have been proposed. Pumping water into a fault zone will increase the hydrostatic pressure, triggering small earthquakes and releasing energy along the fault line. Controlled underground explosions may have the same effect.

8   What are the three types of igneous rocks?

Answer

Firstly, there are silica-poor metal-rich mafic rocks such as basalt ejected from ocean volcanoes. Secondly, there are felsic rocks which are silica-rich, containing metal oxides. Rhyalite, a viscous extrusive type of granite is a good example. Thirdly, there are the intermediate rocks with an average silica content. These are typical of island arcs and young mountain ranges; a good example is andesite.

9   What is a batholith?

Answer

A batholith is a large dome-shaped mass of intrusive rock, often granite. They are formed at great depth within the Earth's crust. They are associated with orogenesis and often form the core of major mountain ranges, such as the Alps. They are the largest form of pluton, being up to hundreds of thousands of square kilometres in extent.

10 How are calderas formed?

Answer

Calderas may form in one of two ways. Some are formed by the collapse of the original volcanic peaks into the empty magma chamber below; repeated eruptions have produced a void unable to bear structurally the weight of the volcano above. Other calderas are produced by violent explosions blowing the top off the erupting volcano and leaving a large, steep-sided crater.

Additional references

Dalziel, I. W. D. (1995) 'Earth before Pangea'. *Scientific American* 272 (1), 38–43.
A well illustrated article that examines a range of evidence supporting the mobility of continental land masses. Links geology, plate margins and biological evidence to provide a 750 million-year sequence of continental spatial movement.

Flanangan, R. (1991) 'Here comes the big one'. *New Scientist* 151 (2039), 36–40.
Discusses the use of seismic waves in forecasting earthquakes. Also suggests the hypothesis that

seismic events are linked and may trigger others in other parts of the world.

Hyndman, R. D. (1995) 'Giant earthquakes of the Pacific Northwest'. *Scientific American* 273 (6), 50–7.
Examines evidence from the geologic record to evaluate earthquake probability. Good synopsis of the earthquake cycle and related plate tectonic processes.

Massonnett, D. (1997) 'Satellite radar inter-ferometry'. *Scientific American* 276 (2), 32–9.
Discusses the use of remote sensors is monitoring change in the Earth's crust. The technology is clearly explained together with applications for future prediction of tectonic hazards.

Stokstad, E. (1996) 'When volcanoes get violent'. *New Scientist* 152 (2053), 32–6.
Looks at the evidence used to predict the type of eruption from volcanoes. Concentrates on magma gas content using examples from around the world.

Taylor, S. R. and McLennan, S. M. (1996) 'The evolution of continental crust'. *Scientific American* 274 (1), 60–5.
A clear explanation of Plate Tectonic Theory. Relates the dynamism of the crustal plate system through time suggesting tectonic cycles.

Wakefield, J. (1997) 'Electric shockers'. *New Scientist* 153 (2071), 34–7.
Outlines new methods for attempting to predict earthquakes by monitoring electric currents in rocks under stress. Good discussion of the transference of laboratory experiments to the real world, and the scrutiny they undergo from the scientific community.

Warren, C. (1997) 'Ice, fire and flood in Iceland'. *Geography Review* 10 (4), 2–6.
Examines the results of volcanic action on the glacial system. These system interactions producing floods are discussed both in terms of hazard potential and as geomorphological agents.

Zhang, Y. (1996) 'Dynamics of carbon dioxide driven lake eruptions'. *Nature* 379, 57–9.
Evidence is presented to produce a hypothesis that the Lake Nyos disaster was hydrodynamically driven and not a result of vulcanism as first thought.

Web site

www.comet.net/earthquake/
A user-friendly site that provides both academic and general media information on earthquakes. Useful information on processes and disaster outcomes of recent earthquakes.

CHAPTER 6
Earth materials

Aims

- To emphasize the importance of the materials in the Earth's crust both in their own right and as essential components of the global biogeochemical cycles.

- To examine the various roles they play in the soil and biological systems of the Earth.

- To assess the processes and mechanisms which act on the Earth's materials, the products of these actions and the effects on landforms. The redistribution of material is viewed as the key concept.

Key-point summary

- *Minerals* are the basic components of all rocks in their original state or the products of interaction with other system processes, e.g. weathering down to produce inorganic soil components.

- Minerals are relatively consistent in their chemical make-up. As part of a system they, therefore, exhibit stable properties.

- The structure of most minerals is of a crystalline nature. The two key variables of this structure are the size of crystals and the strength of bonding between them. These properties will determine the rates of weathering.

- Rocks may be classified according to the directness of their link to the internal rock-forming material – magma. *Igneous rocks (primary)* are formed by the cooling and solidifying of molten magma. They are constantly being created, and destroyed as part of the crustal conveyor belt powering plate tectonics. It should be noted that the *subduction zones* at the margins of some crustal plates also destroy sedimentary and metamorphic rocks. *Sedimentary rocks (secondary)* are derived from the weathering and deposition of other pre-existing rocks. They are a product of surface processes related to weathering systems. *Metamorphic rocks (tertiary)* are derived from igneous and sedimentary rocks altered by natural geological processes.

- Igneous, sedimentary and metamorphic rocks have particular characteristics which determine how they react with, and contribute to, other environmental systems

- *Acid* igneous rocks often weather into coarse-textured, infertile soils. *Basic* igneous rocks weather down into more fertile soils, where mineral elements are more freely available for the biotic system. *Clastic* sedimentary rocks can be discriminated by particle size. They are formed by deposition, mainly in water environments, where particle size has an influence on spatial distribution and rate of deposition within geosynclinal depressions in the global relief system. Some sedimentary rocks link directly to the biological system rather than the tectonic system. Limestone derived from coral is a common example. Here the ocean system internalizes the rock-forming process.

- All rocks are stages in a continuous cycle where crustal material is being refashioned, redistributed and recycled by the processes associated with the tectonic and denudational systems. This cycle operates at a much slower rate than the biogeochemical and global water cycles.

- *Denudation* processes lower surface relief, in effect operating in the opposite manner to the sub-surface processes which overall tend to raise the surface (*diastrophism*). These operate at the *global scale* providing a dynamic system which tends to equilibrium at this scale. At *smaller scales* the balance between the two may be unequal, leading to either raising or lowering of the surface. Overall system control is maintained by a *feedback loop* where raised relief increases the rate of weathering by increased gravitational force (height) and often greater climatic variability.

- Changes in relief significantly influence other surface and atmospheric environmental systems.

- Weathered materials may be removed through the surface system by *erosion* or may remain in situ where through time they become a component of soils.

- Human activities affect the atmospheric system. Inputs of air pollution increase the rates of weathering, particularly noticeable on building stone.

- Weathering comprises two main types – *physical disintegration* and *chemical decomposition*. The physical breakdown of rock may be due to interaction from the climatic system (*frost, insolation* and *salt* weathering) or the biological system (e.g. *root wedging*). Decomposition of rock is fuelled by chemical reactions with components of the atmospheric and water systems. These components may be increased artificially by human activity, e.g. carbon dioxide from fossil fuel burning, or by interaction with the biological system – the absorption and release of organic substances to form weak acids.

- *Erosion* processes are the flows in the denudational system. Transporting agents such as running water, moving ice or wind move material. They are an inherent part of the way surface systems operate.

- Human land use activities have an important effect on the water transporting mechanism within the hydrological system. The soil system may be protected or laid bare to direct precipitation input by human-induced vegetational change, mainly forest clearance and agriculture practices.

- As with other processes, erosion varies from place to place, controlled by the mosaic of climate, vegetation, land use, topography, geology and soil. Thus, at the global scale, they are all interdependent systems with the balance of them at smaller scales producing unique process characteristics locally.

- *Mass movement* of weathered material downslope utilizes gravity as the main agent. The range of processes and their speed of motion (*creep, flow,*

slide, slump and *fall*) are dependent on the strength of their links with other agents, the lubrication of weathered material by water being the main control. Mass movement illustrates a system (*slope*) that is unstable.

- The rates of movement and inception of instability in a slope system may also be influenced by human action such as cultivation and leisure activities.

- The threshold of instability may be crossed by action in other physical systems adjacent to the slope system. Mudflows, avalanches and landslides have been triggered by earthquake activity. Global systems, affected by human activity, may also be related. Climate change resulting in increased rainfall intensities has been recorded since the 1920s.

- As with other hazards the spatial mapping of mass movements and subsequent land-use zonation in relation to them has reduced the risk to people and property.

Main learning hurdles

Erosion, weathering and denudation

Often confusion arises in the use of these terms, with the student often interchanging them. This is especially true of erosion and weathering. The instructor should therefore be certain that the student understands the basic weathering and erosion processes, a key element being the transport mechanisms involved in erosion and subsequent deposition. Denudation may then be viewed as the total system outcome of these processes, including mass movement. Another area that sometimes causes problems is the concept of differential weathering within the same rock type. Students readily accept different rates of weathering between rock types but have difficulty at the single rock-type scale. It is a useful exercise to bring in a variety of rock samples and ask them observational questions in relation to colour, texture and mineral mix. They should be able to see clear differences between the rocks with a more homogeneous make-up and those without.

The equilibrium of slopes

Students are sometimes confused when deciding if a slope system is in dynamic or steady-state equilibrium. The problem is due to the long timescale involved in changing slope system conditions. Over a short timespan of a few years it is difficult, in general, to detect change. However, over a longer timespan (centuries) there may be evidence to suggest a steady-state maintained by negative feedback loops. Yet, over thousands of years the system might exhibit a generally decreasing gradient with short-term fluctuations around this trend, thus exhibiting dynamic equilibrium. As with other systems the timescale of examination (evidence) needs emphasizing.

Forces acting on slope particles

For students with a non-scientific background it would be useful to spend some time covering shear stress, shear strength and friction. It is important they realize that gravity acts vertically and not downslope.

Key terms

Acid and basic rocks; avalanche; base level; carbonation; chemical decomposition; clastic; creep; denudation; diagenesis; diastrophism; elements; evaporites; fall; flow; frost weathering; ground frost; groundwater; gully development; hydrolysis; igneous; insolation weathering; lithification; metamorphic; minerals; organic processes; oxidation; physical disintegration; precipitates; rainsplash; salt weathering; sedimentary; shear strength; shear stress; slide; slope erosion; slump; soil water; stratification; undercutting; weathering.

Issues for group discussion

Discuss the contention that slopes are in equilibrium

The discussion should emphasize the different timescales involved. Reference to Chapter 2 is useful. The discussion may usefully broaden out to consider natural global system states, modelled by isostasy and eustasy, and also an assessment of human systems.

Discuss the distinct forms produced by different weathering processes

Students should consider the sections on weathering processes given in the chapter as a starting point.

Doornkamp and Ibrahim (1990) give a focused case study and may be used to open out the discussion into consideration of the importance of climatic variables. Weather patterns, as well as system change due to human interference, are also considered by Meierding (1993). By the end of the discussion form and process should be appreciated, possibly tied in with differential weathering (see major learning hurdles). Finally, a generalized regional pattern related to atmosphere variables should be achieved.

Discuss the avalanche hazard in relation to human activity

The discussion should revise a range of activities in relation to mountain environments, including agriculture and recreation/tourism. These should be evaluated regarding their value and the ability to mitigate the hazard. Reference to developed and developing countries would be useful, e.g. Alpine versus Andean experience. Reference to Smith (1988) should allow a discussion of why there is an increase in deaths from avalanches in certain areas.

Selected reading

Doornkamp, J. C. and Ibrahim, H. A. M. (1990) 'Salt Weathering'. *Progress in Physical Geography* 14(3), 335–48.
A comprehensively referenced article on salt weathering, its mechanisms, effects and distribution. Links natural system status with hazard potential to provide an application for planning management.

Meierding, T. C. (1993) 'Marble tombstone weathering and air pollution in North America'. *Annals of the Association of American Geographers* 83 (4), 568–88.
The use of empirical weathering evidence clearly demonstrates the major effect air pollution has in enhancing natural atmospheric weathering processes. Quantifies rates of surface recession and links these spatially with weather patterns.

Smith, K. (1988) 'Avalanche hazards: The rising death toll'. *Geography* 73 (3), 157–8.
A short article demonstrating quantitatively the avalanche hazard as mainly affecting economically developed countries. It links human economic systems with climate and topographical systems at the regional scale.

Textbooks

Bridges, E. M. (1990) *World Geomorphology.* Cambridge University Press: Cambridge.
A very useful examination of the action of surface processes at the continental scale.

Cooke, R. U. and Doornkamp, J. C. (1990) *Geomorphology and Environmental Management,* 2nd edn. Oxford University Press: Oxford.
A comprehensive introduction to Earth surface processes and their interaction with human management systems.

Gerrard, A. J. (1990) *Mountain Environments.* Belhaven: London.
A range of resources is reviewed in these discrete environments. Good section in relation to slopes, materials and associated hazards.

Goudie, A. (1995) *The Changing Earth: Rates of Geomorphological Processes.* Blackwell: Oxford.
This text considers the present state of knowledge of Earth surface processes and morphological change. It provides topical system links with the hydrosphere and atmosphere. Integration of the upper lithosphere is provided by an account of tectonics.

Parsons, A. J. (1988) *Hillslope Form.* Routledge: London.
Extremely useful in explaining process theory. Clearly links knowledge of the slope system to classifying associated hazard events.

Selby, M. J. (1993) *Hillslope Materials and Processes,* 2nd edn. Longman: London.
This book has an introductory multidisciplinary approach suitable to the variety of students studying the environment. The slope system is extensively examined in relation to erosional and depositional processes as well as its fundamental interactions with the geological and soil systems.

Essay questions

1 Evaluate the controls on the pattern of slope evolution.
2 Are concave-convex slopes a landform exhibiting equilibrium?
3 Discuss the importance of chemical weathering processes.
4 Evaluate the hazard potential of landslides at the global scale.

5 How might human actions promote mass movement?
6 Discuss the factors which influence the resistance of a rock to weathering.
7 Using a systems approach, describe the processes, stores and control mechanisms that characterize the slope system.
8 Discuss the importance of scale in geomorphological explanation.
9 Evaluate the factors causing rapid mass movement of weathered material on slopes.
10 Outline the rock cycle and illustrate the interrelationships between rock types and the processes that link them.

Multiple-choice questions

Choose the best answer for each of the following questions.

1 Exposed rocks have an oxygen content of about:
   (a) 7 per cent
   (b) 27 per cent
   (c) 47 per cent   *
   (d) 67 per cent

2 The scale used to measure mineral hardness is the:
   (a) Mineral Hardness Scale
   (b) Moh Resistance Scale
   (c) Moh Hardness Scale   *
   (d) Moho Hardness Scale

3 Basalt is:
   (a) the most common extrusive igneous rock   *
   (b) the least common extrusive igneous rock
   (c) the most common intrusive igneous rock
   (d) none of the above

4 How much of the Earth's surface is exposed granite?
   (a) 5 per cent
   (b) 15 per cent   *
   (c) 25 per cent
   (d) 50 per cent

5 The phi scale expresses:
   (a) the hardness of a mineral
   (b) the distance of a particle from source
   (c) the size of individual particles   *
   (d) the angularity of a particle

6 The most common sedimentary rock is:
   (a) slate

(b) shale   *
(c) chalk
(d) sandstone

7 Metamorphism may be caused by:
(a) pressure
(b) heat
(c) chemical reactions
(d) (a), (b) and (c)   *

8 Voids or pore spaces are found in which rocks:
(a) sedimentary   *
(b) metamorphic
(c) igneous
(d) (a), (b) and (c)

9 Erosion of the world's continents results in the loss per year of:
(a) 10 million hectares of tropical forest
(b) 10 million tonnes of sediment   *
(c) 10 billion litres of water
(d) 10 billion hectares of marine habitat

10 Rivers account for approximately how much of continental erosion:
(a) 10 per cent
(b) 33 per cent
(c) 60 per cent
(d) 95 per cent   *

11 Human activities have contributed to serious land degradation in the order of:
(a) 100 thousand square km
(b) 1 million square km
(c) 11 million square km   *
(d) 100 million square km

12 Which of the following is not a type of mass movement:
(a) saltation   *
(b) slide
(c) fall
(d) creep

13 Solifluction is very widespread in:
(a) temperate environments
(b) tundra environments   *
(c) tropical environments
(d) desert environments

14 Which of the following mass movements is usually slow:
(a) avalanche
(b) mudflow

(c) rockfall
(d) solifluction   *

15 The proportion of voids in a rock is known as:
(a) porosity   *
(b) specific gravity
(c) density
(d) volume

Figure questions

1 Figure 6.11 depicts the relationship between land-use change and sediment yield in the Piedmont region of Maryland. Answer the following questions.
(a) Describe the relationship between the land-use and sediment yield.
(b) What processes account for this trend of sediment yield through time?

Answers

(a) Under a natural forest sediment yield is very low. As more land is taken over for cropping sediment yields rise. With a reduction in cropping area, promoting secondary woodland and a conversion to grazing, sediment yields fall but stay at a level higher than under natural forest. The short period of construction promotes a peak of sediment yield three times higher than anything previously experienced. Under urbanization the sediment yield returns to its previous low.

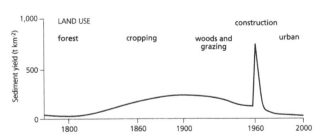

Figure 6.11 *Relationship between land-use change and sediment yield in the Piedmont region of Maryland. Sediment yield increased when forest was cleared and replaced by crops, and declined slightly when secondary woodland growth protected soils. A short phase of building activity in the drainage basin caused a sharp rise in erosion, followed by reduced yield in the urbanised basin. After Figure 24.23 in White, I.D., D.N. Mottershead and S.J. Harrison (1984)* Environmental systems. *George Allen & Unwin, London*

(b) Given the time-scale of the graph and the correlation with land-use types, significant changes in precipitation input may be effectively discounted. The processes accounting for the trend shown operate in relation to rainfall impact, the exposure of soil, and the ability of the surface to promote overland flow. Under forest the tree canopy reduces rainfall impact, thus little soil material will be physically disaggregated. Also precipitation will transfer to the surface via the vegetation at a slower rate, reducing the opportunity for overland flow (infiltration capacity is not exceeded). Under cropping the soil surface is not protected for large parts of the year and human activity will break up the soil. Except when under a mature crop rainfall interception will be less, thus rainfall impact and overland flow transporting sediment will increase. Grazing reduces the human interference and the grass cover protects the soil surface better than intermediate cropping. The capacity for overland flow will still be there but the supply of sediment will be less. Construction causes great physical break up of the soil, increasing the supply. The soil will also be unprotected and compacted, reducing infiltration capacity and increasing overland flow. Under urbanization overland flow will be promoted by the impermeable surface but the supply of sediment will be drastically reduced as the soil is not exposed to precipitation or surface flow.

2 Figure 6.16 illustrates rapid mass movement processes. Answer the following questions.
   (a) List the mass movements in order of speed from fastest to slowest.
   (b) Briefly describe how each mass movement is instigated and explain in general terms.

Answers

(a) Falls, flows, slumps and slides.

(b) Falls are associated with hard rock slopes of steep angle. Weathering, such as freeze-thaw processes, produces failure in the rock which falls rapidly under gravity. Flows occur in saturated materials. Here the material acts like a fluid and flows rapidly downhill with friction reduced by lubrication. Slumped material has defined planes of weakness due to the distribution of stresses in the material. As they rotate down a concave plane they are quite rapid due to high back slope angle but frictional forces and reduced lower slope angle mean they are slower than falls and flows overall. Slides may be triggered when a balance at the angle of repose is disrupted by vibration or rainfall. A generally lower slope angle for these slides means they are the least rapid.

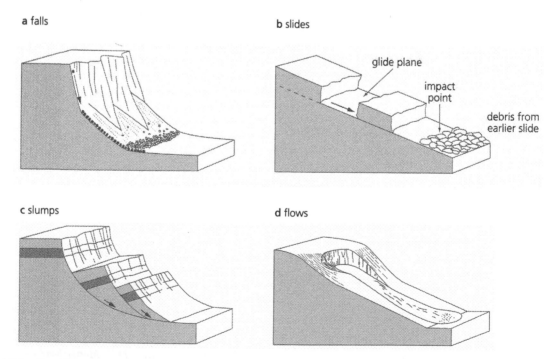

Figure 6.16 *Rapid mass movement processes. Falls (a), slides (b), slumps (c) and flows (d) operate at different speeds and produce different landscapes. Source unknown*

Short-answer questions

1 What are the main properties of igneous rocks?

Answer

Igneous rocks contain no fossils. Their appearance is quite uniform, usually with a lack of stratification. Structurally they contain no voids, consisting entirely of interlocking minerals in a crystalline form.

2 What types of materials form sedimentary rocks?

Answer

Rock fragments from the land surface may be cemented together to produce clastic sedimentary rocks. Organic material and chemicals deposited in water environments both produce sedimentary rocks.

3 Define lithification.

Answer

Lithification or diagenesis is the process of compaction (compression by overlying material) and cementation (mineral bonding) of unconsolidated sediments to form sedimentary rock.

4 Describe the formation of fossil fuels.

Answer

The main fossil fuels of coal, oil and natural gas are formed by the decomposition of organic material under anaerobic conditions. Coal, by the process of lithification, forms a sedimentary rock. Oil migrates as a viscous mixture into other strata as compaction of the source rocks intensifies. Natural gas is trapped in a gaseous state in underground rock reservoirs.

5 List the changes involved in metamorphism.

Answer

Metamorphism may involve increases in pressure, for example, tectonic plate convergence; increases in heat, such as adjacency to magma; and chemical reactions related to gas and liquid injection.

6 What is weathering?

Answer

Weathering is the process of disintegration and decomposition of a rock in situ. The rate of weathering is dependent on the chemical and physical nature of the rock and climate.

7 Outline the main principles of weathering.

Answer

All rocks at or near the surface are subjected to weathering. Different rocks weather at different rates due to their individual characteristics. Different climate regimes will weather the same rocks by different processes and at different rates. Physical faults in a rock (fissures) will enhance the speed of weathering.

8 Describe the controls affecting chemical weathering.

Answer

The principle controls on chemical weathering efficiency are climatic. Generally water is required for chemical reaction so the more humid the climate the more favourable the conditions. Heat usually enhances the speed of chemical reactions so a warmer climate will favour chemical weathering. The final control is the chemical and physical nature of the rock, which acts as an ultimate control to reaction with both water and heat.

9 Outline the main forces acting on a particle.

Answer

Shear stress is the downward force of gravity. On a slope it promotes downslope movement. If all variables are equal it has a positive relationship with slope angle. Shear strength is the opposing force resisting movement, such as frictional resistance and internal cohesiveness. The particle might also be physically bound by outside forces such as plant roots.

10 List the sequence of events in the Vaiont Dam disaster, 1963.

Answer

The construction of a high dam, with a reservoir behind, across a narrow valley in 1960; the reservoir was filled by 1963; a rock slide took less than a minute to fill a large part of the reservoir to a height 150 metres above the dam; the displaced water almost immediately overtopped the dam by 100 metres; two minutes later this passed down the valley as a 70 metre wave; the valley floor infrastructure was flattened and 2,600 people were drowned; within 15 minutes of the hillslope failure the flood waters had gone down.

Additional references

Higgitt, D. (1995) 'Earth surface processes'. *Geography Review* 9 (2), 13–15.
A concise introduction to the forces behind Earth surface dynamism. Slope instability is explained and an applied focus is given by discussion of soil erosion and strategies for management from a multi-disciplinary viewpoint.

Holiday, A. (1997) 'Managing the Wessex coast: The impact of mass-movement on the Furzy cliff coastline'. *Geography Review* 10 (5), 32–5.
A discrete case study of the interaction of erosional processes and geology. The study considers the nature of the problem in relation to hazard management solutions. An applied aspect is introduced via cost–benefit analysis of proposed engineering works.

Jakob, M. (1996) 'Debris flows: Curse of the mountains'. *Geography Review* 9 (5), 34–6.
An overview of the factors responsible for these natural hazards. Event frequency and magnitude are considered as predictive data. The article reinforces that knowledge of a number of interacting systems accrued from a multidisciplinary approach is needed for a fuller explanation of these events.

Pinter, N. and Brandon, M. T. (1997) 'How erosion builds mountains'. *Scientific American* 276 (4), 60–5.
Investigates the genesis of mountains by linking crustal tectonic processes with wind and water erosion. Useful in examining the interactions between the lithosphere and atmosphere as environmental systems that produce surface topography. The various scales of system processes are also discussed.

Young, P. (1996) 'Mouldering monuments'. *New Scientist* 152 (2054), 36–8.
Examples of micro-organisms weathering buildings. Illustrates a clear link to the role of air pollution in fuelling the problem.

Web site

earth.agu.org/kosmos/homepage
This site can be a little slow to access. However, it contains a wealth of links to the earth sciences.

Aims

- To emphasize the importance of the atmosphere in sustaining planetary life, and how susceptible it is as a system to detrimental inputs that might reduce this ability.

- To examine the important processes and transfers that take place within the atmosphere and between it and other environmental systems.

- To evaluate the atmosphere in relation to other systems, indicating effects on natural systems and in turn human systems.

- To explore human activities that threaten the integrity of the atmospheric system. With special emphasis on its finely balanced dynamic and responsive nature.

Key-point summary

- The atmosphere is a natural, energy-driven, open system powered by solar radiation. This system determines our *weather* and *climate*.

- The balance of gases in the atmosphere is crucial to life on Earth. Human impacts, through *air pollution* and *land-use change* can alter this balance. The scale of these impacts has increased rapidly through time and space during the twentieth century.

- As the atmosphere is all embracing and responds rapidly to change then local inputs to the system soon manifest themselves as large-scale problems, often global in nature and impact.

- The high mobility of its constituents means that whilst the global atmospheric system may be in a stable state, local variations occur within the atmosphere and at the interface with surface systems. Variations also occur through time.

- The atmosphere comprises *liquids* and *solids* as well as *gases*.

- The relative importance of atmospheric constituents to other environmental systems is not proportional to their absolute amount. Of the gaseous constituents, *carbon dioxide*, needed by plants for use in *photosynthesis*, is crucial but only accounts for less than one thousandth of the total volume of atmospheric gases. However, an increase in the volume of these low-volume naturally present gases such as carbon dioxide and *methane* due to inputs from human systems upsets the atmospheric system equilibrium. Gases not present in the natural atmosphere but produced by, or instigated by, human economic systems have the same disproportionate deleterious effects.

- Water vapour plays a large role in determining weather conditions and temporal changes. It is essentially confined to the lower 12 km of the atmosphere.

- The majority of solids in the atmosphere originate from natural systems. Air pollution from human activities is on the increase and contributing more and more particulate matter to the atmosphere.

- Weather is determined by conditions within the atmospheric system. Weather patterns and conditions are mostly confined to the lowest 16 km, though air movements up to 30 km will affect these.

- There is a vertical differentiation in the atmosphere: the lower atmosphere (*troposphere*); the middle atmosphere (*stratosphere*) and the upper atmosphere (*mesosphere and thermosphere*).

- The troposphere serves as a link between environmental flows and systems on the land and ocean below and the stratosphere above. Being adjacent to the Earth's surface it is in direct receipt of pollutants from human activities. Some of these find their way through the troposphere to the stratosphere, e.g. CFCs affecting the ozone layer.

- As the regulator between incoming solar radiation (nature and amount) and the surface of the planet which contains life, any effect on the atmosphere can have dire consequences. Importantly, due to the all embracing nature of the atmospheric system, these are often of global significance in both cause and effect.

- The relationship between the solar life source and the Earth's species is complex and is dependent on a variety of large-scale systems. Our position in the solar system gives the Earth a biological window of opportunity related to the heating effect of the Sun. The Earth's system of motions regulates how long regions receive direct solar radiation both seasonally and diurnally. Atmospheric depth, in line with the direction of solar radiation, regulates the amount received along a meridian during any day – a regional variation controlled by the Earth's shape.

- The atmospheric system acts as a series of filters with different system components removing, or reducing, certain wavelengths of the electro-magnetic spectrum. In addition scattering and reflectance by water vapour and particulates within the atmosphere reduces the overall amount of incoming solar radiation.

- Reflectance both within the atmosphere and on the Earth's surface is variable and expressed by a reflection coefficient or *albedo*. Albedo is a regulator in the biosphere system and a control on climate stability. Human-induced change affecting the surface/troposphere albedo produces feed-back into the climate system potentially of global significance.

- Our scientific certainty about how energy flows and transfers within the atmosphere is limited; especially related to change in system integrity related to non-natural inputs. Thus, *simulation models* predicting the future course of global atmospheric problems are variable, being based on imperfect knowledge of an extremely complex system. This uncertainty is amplified by the inter-action of the atmospheric system with surface systems, especially the oceanic circulatory system.

- The *greenhouse effect* is a natural process of the

biosphere powering the biogeographical cycles and sustaining life on Earth. Only certain atmos-pheric gases absorb the radiated long-wave energy from the Earth, and these act as short-term stores in the system.

- The Sun provides our energy resources. Over long time-scales the biological products of photo-synthesis are stored in the lithosphere as *fossil fuels*. Contemporary solar energy is also exploited directly by *solar furnaces* or indirectly by *solar cells* which convert the solar energy to electricity.

- Air pollution reflects the influence of the human economic system on the atmosphere. These influ-ences may be either deliberate (waste disposal) or accidental (industrial disaster). The intimate inter-action of the atmosphere with the biosphere means that atmospheric system pollutants have significant outcomes for many species, including humans. The dynamism of the atmosphere means these outcomes spread spatially and become, through time, global problems. This is true of *global warming*, *ozone depletion* and *acid rain*. Air pollution may be concentrated locally, such as urban air pollution, but it usually has a global causal distribution, e.g. urbanization and traffic. As air pollution is perceived as a global problem it is part of international environmental policy and features prominently in *Agenda 21*.

- Whilst the most important global air pollution problems are widespread both in cause and effect, individual incidents may have devastating effects regionally (*Chernobyl*) and locally (*Bhopal*). They are also extremely newsworthy and graphically illustrate human impact on the environment. One important aspect of these industrial accidents is that they introduce substances not normally found in the atmospheric system.

- Discrete air pollution effects are found at various levels within the atmosphere. At low level, urban air pollution illustrates the synergistic nature of combining pollutants with the alteration of local climate. These two system impacts associated with urbanization produce from an input of car exhaust gases (particularly *nitrogen oxide*) and a local microclimate that favours the persistence of *smog* a significant local environmental health problem. At higher levels within the stratosphere the natural 'ozone layer' is depleted by inputs of *CFCs*. As the protective ozone layer surrounds the Earth, this is a global problem. Response to it was similarly global with major international protocols agreed and serious attempts made to halt the problem even though scientific knowledge was not conclusive. This proactive stance illustrates the *precautionary principle* when addressing environmental problems.

- Acid rain demonstrates the uneven (and unfair) distribution of some atmospheric pollution. The atmospheric system moves pollutants away from source areas via the wind circulation sub-system and hydrological store (clouds) to be deposited elsewhere. This is a good example of *trans-frontier* air pollution, with some countries being net *exporters* of the problem and other countries being net *importers*. However, there is some conjecture over the actual system response to acidic pollutant *critical loads*. This time, without absolute scientific knowledge, response to tackling the problem has been slow and firm international policy commitment has stagnated at a reduction of *30 per cent*. A *'wait and see'* attitude was evident in many countries.

- Human impact on environmental systems is not confined to damaging them or introducing unnatural materials into their system operation. Enhancement of a systems function may cause threatening environmental problems. The increased input of greenhouse gases into the atmosphere has reinforced the 'greenhouse effect', appearing to cause global warming. The problem is magnified by the human effect of both increasing input of gases, from fossil fuel burning and ranching, and reducing output of carbon dioxide by tropical forest felling. Again this is a global problem needing global solutions based on evidence from many disciplines.

- The implications of global warming demonstrate the symbiotic relationship of the Earth's major environmental systems and our lack of absolute knowledge of them. There is still debate over the balance of causes between natural climate cycles and air pollution. Future predicted effects are hard to quantify as the rate of temperature rise is still a matter of conjecture. However, what is likely to be affected illustrates the interdependency of the Earth's systems. Increased global temperatures will cause the oceanic store to expand with an ensuing rise in sea level. Climate and weather will change with subsequent effect on surface biotic systems, both natural and anthropogenic. As a major global problem, global warming requires global solutions. However, international policy agreement is hampered by the different framing of the problem by the economically developed world and economically less developed world.

Major learning hurdles

Jet streams

Students sometimes have difficulty in conceptualizing jet streams, particularly in relation to general wind patterns covered in the next chapter. It is probably best to review them with Figure 7.1 in relation to temperature and altitude and stress their global distribution and direction. Coriolis force may be brought into the discussion, but at this general level suggested causes should be secondary to their effects on climate.

Radiation in the atmosphere

Some time should be spent on explaining the nature of the electromagnetic spectrum, focusing on wavelengths associated with heat and light energy. This is fundamental to the students' appreciation of radiation flows in the atmospheric system and associated disruptive impacts on these flows. It will also provide a foundation for the understanding of our monitoring of the environment by a range of remote sensors.

Atmospheric chemistry

Diagrams illustrating air pollutant reactions with water, oxygen and carbon dioxide in the atmosphere sometimes deflect students from the important environmental outcomes. It is important to stress, as with biogeographical cycles, that this is fundamental knowledge required to understand our environment and its reaction to change. Therefore, some time should be spent in working through the main chemical formulae and equations so that students may appreciate the dynamism of the atmospheric system and gain the most from diagrams in this and other sources.

Key terms

Absorption; acid rain; albedo; Bhopal; Brundtland Commission; 'business as usual' scenario; carbon dioxide; catalytic converters; Chernobyl; chloro-flourocarbons (CFCs); critical thresholds; gases; international environmental problems; mesosphere; methane; Montreal Protocol; natural and enhanced greenhouse effect; nitrogen oxide; 'no regrets policy'; photochemical smog; photo-oxidants;

photovoltaic cells; precautionary principle; reflection; sea-level rise; solar collectors; standard lapse rate; Stefan-Boltzmann equation; stratosphere; sulphur dioxide; sunlight; thermosphere; tropopause; troposphere; water vapóur; Wien's law.

Issues for group discussion

Discuss the contention that 'dilution is the solution to pollution'

This discussion should focus on the acid rain problem. Students should investigate the history of pollution disposal by chimneys and evaluate the contention from both the local and wider perspective. Park (1988) clearly illustrates the basic system action and the associated spatial implications. If the discussion rejects the contention, then reductions in pollution can be usefully explored regarding who will benefit from this action. This discussion may usefully be referred to with regards to marine pollution as well.

Discuss the environment at risk from global warming

Warrick and Farmer (1990) are essential reading for this discussion. The students should particularly focus on three key issues. Firstly, the range of future scenarios based on different rates of warming – the worst and best cases. Secondly, the implications of sea-level rise for coastal environments. Thirdly, the more complex issue of shifting climate belts affecting habitat and agricultural distributions. This last point should emphasize the migrational aspects and associated political problems. Recent African case studies could be usefully included.

Selected reading

Oliver, H. and Oliver, S. (1993) 'The ins and outs of sunshine'. *Geography Review* 7 (1), 2–6.
An introductory account of the importance of solar energy in powering the Earth's systems. Inputs and outputs are clearly explained with good illustration.

Park, C. C. (1988) 'Acid rain: Trans-frontier air pollution'. *Geography Review* 2 (1), 20–5.
A systems review of the nature and components of the acid rain problem. The international multi-disciplinary discussion highlights both the scientific and political complexities of the issue.

Perry, A. H. (1985) 'The nuclear winter controversy'. *Progress in Physical Geography* 9 (1), 76–81.
A spatial evaluation of the dynamics of atmospheric systems in dispersing nuclear atmospheric pollution. The fragility of the climate system threshold to a rapid short-term material input is emphasized by the interaction between surface and atmosphere. Perhaps a little dated in today's political climate but very useful for process comparison with natural events such as large volcanic eruptions.

Warrick, R. and Farmer, G. (1990) 'The greenhouse effect. Climate change and rising sea level: Implications for development'. *Transactions, Institute of British Geographers* 15 (1), 5–20.
A comprehensive account of causes, system effects and the human outcomes of likely climatic change resulting from the enhanced greenhouse effect. Sound emphasis on spatial patterns of change and impact areas, with a reinforcement of the global generation of inputs and the interaction of major environmental systems to these.

Textbooks

Adgar, W. N. and Brown, K. (1994) *Land Use and the Causes of Global Warming*. Wiley: Chichester.
Presents a range of case studies of land-use effects on atmospheric composition and functioning. Useful material on the relationship between forest ecosystem exploitation and global warming.

Bradby, H. (ed.) (1990) *Dirty Words: Writings on the History and Culture of Pollution*. Earthscan: London.
Contains a very useful section on air pollution, waste products and climatic effects. Clearly informs the student of the historical change in valuing our environment.

Bridgman, H. (1990) *Global Air Pollution*. Belhaven: London.
Considers a range of air pollution inputs. A very useful urban scale chapter illustrates the interference of human systems on atmospheric processes with subsequent detrimental impacts. Also contains a key section on the veracity of forecasting future scenarios.

Crowley, T. J. and North, G. (1991) *Paleoclimatology*. Oxford University Press: New York.
A detailed presentation of past climatic evidence and

its use for predicting impacts of global warming. Clearly demonstrates cyclical patterns at different time-scales with solid links to system concepts.

Elliott, D. (1997) *Energy, Society and Environment: Technology for a Sustainable Future*. Routledge: London.
A well-structured, introductory text on societies' use of energy and its primary effects on the atmospheric system. Usefully considers the technological optimist approach versus the greener sustainable argument.

Elsom, D. (1992) *Atmospheric Pollution: A Global Problem*, 2nd edn. Blackwell: Oxford.
The global perspective is clearly explained by a synthesis of point source pollutants at the local scale providing wider spatial scale problems. Environmental pathways and policy pertaining to movements of pollutants between the atmosphere and the Earth's surface are evaluated. Good range of national case studies for class discussion.

Elsom, D. (1996) *Smog Alert: Managing Urban Air Quality*. Earthscan: London.
An interesting focus, at the smaller scale, of urban air pollution. The problems are clearly presented and suggested strategies are explored. A good range of case studies is drawn on from around the world. Environmental monitoring and multidisciplinary management is encouraged.

Turco, R. P. (1996) *Earth under Siege: From Air Pollution to Global Change*. Oxford University Press: Oxford.
An introductory text concerning the use of scientific evidence and concepts in defining a range of environmental problems. These problems are usefully illustrated at a variety of scales, interlinking the local through to global. Useful explanation of fundamental earth science principles underpin the many illustrations and examples.

Watkins, K. (1995) *The Oxfam Poverty Report*. Oxfam: Oxford.
Contains an appraisal of the global effects of climate change. Detailed focus on the economically developing countries with suggestions for sustainable management.

Whitelegg, J. (1993) *Transport for a Sustainable Future*. Belhaven: London.
This readable text includes chapters on the health effects of local air pollution and the implications of global warming. The inputs from transport systems are thoroughly explored and sustainable future policies suggested.

Whyte, I. D. (1995) *Climatic Change and Human Society*. Arnold: London.
The history of human inputs of air pollution is discussed. Climatic reconstruction and future prediction is related to global warming.

Woodwell, G. M. and Mackenzie, F. T. (1995) *Biotic Feedback in the Global Climatic System: Will the Warming Feed the Warming?* Oxford University Press: Oxford.
An examination of the global carbon cycle and its response to global warming. Potential biotic feedback is identified as a mechanism for intensifying the problem of climatic change.

Essay questions

1 'Clouds and water vapour are the main regulations of the Earth's heat system.' Discuss.
2 Illustrate the importance of the atmosphere in the global carbon cycle.
3 Outline and account for the vertical distribution of temperature change in the atmosphere.
4 How does carbon dioxide influence the Earth's radiation balance?
5 With reference to environmental systems' response to acid precipitation input discuss the concept of critical loads.
6 Why should governments spend money on curbing 'greenhouse gas' emissions?
7 What will be the benefits of reducing sulphur emissions from power stations in Western Europe?
8 Outline a global forest resource policy that might help alleviate global warming.
9 What are the future global climatic changes expected to result from the enhanced greenhouse effect? Assess the possible impact of these changes on society.
10 Explain the main causes of atmospheric pollution.

Multiple-choice questions

Choose the best answer for each of the following questions.

1 The most common gas in the troposphere and lower atmosphere is:
   (a) argon

(b) carbon dioxide
(c) nitrogen   *
(d) oxygen

2 The amount of dust deposition from the atmosphere per day generated by human activity is:
(a) 25 per cent   *
(b) 15 per cent
(c) 5 per cent
(d) 1 per cent

3 The lower atmosphere is comprised of the:
(a) troposphere and stratosphere
(b) stratosphere
(c) troposphere   *
(d) mesosphere and thermosphere

4 What has the highest albedo value?
(a) concrete
(b) desert
(c) cumulonimbus cloud
(d) fresh snow   *

5 The temperature of the Earth's surface if there were no atmosphere may be calculated by:
(a) Wien's Law
(b) sedimentary evidence
(c) Bosman's Law
(d) the Stefan-Boltzmann equation   *

6 Energy wavelengths radiated by the Earth are:
(a) shorter than those of the Sun
(b) longer than those of the Sun   *
(c) the same as those of the Sun
(d) (a), (b) and (c)

7 The main contributor of nitrogen dioxide to local air pollution is:
(a) space heating
(b) coal burning
(c) traffic   *
(d) iron and steel production

8 Why has the prospect of a nuclear winter diminished?
(a) ice caps are melting
(b) the Earth is moving nearer the Sun
(c) the Cold War has ended   *
(d) because of global warming

9 Britain's first air pollution inspector, Robert Smith, found evidence of acid fallout in:
(a) 1792
(b) 1852   *

(c) 1902
(d) 1952

10 The most effective strategy for reducing sulphur dioxide in the atmosphere is to:
(a) lime lakes
(b) use learn-burn technology
(c) plant more trees
(d) reduce emissions from power stations   *

11 WMO stands for the:
(a) World Methane Organization
(b) Water Modelling Office
(c) World Meteorological Office
(d) World Meteorological Organization   *

12 Global sources of methane as a product of human activity principally come from:
(a) livestock   *
(b) rice fields
(c) coal burning
(d) solid waste

Figure questions

1   Figure 7.4 is a generalized diagram of the fate of incoming solar radiation. Answer the following questions.
(a) What would be the major pathways of radiation at night?
(b) Suggest three differences of energy exchanges between the atmosphere and an ocean surface compared to a continental surface.
(c) Outline three ways in which human activity modifies this energy system.

Answers

(a) Re-radiated heat will flow from the soil heat store built up during the day. This long-wave radiation will be transferred to the atmospheric store. In the upper atmosphere some will be lost to space.

(b) As the ocean surface is more homogeneous, there is less differential heating. The open water surface of the ocean would promote greater evaporation outputs. The ocean would generally reflect less solar radiation and thus absorption rates would be greater.

(c) Land-use change will alter the surface albedo. Air

(a) In general terms why is the percentage of methane emissions from the USA so much higher than other countries?
(b) What are the environmental effects of methane emissions?
(c) How might methane emissions be reduced?

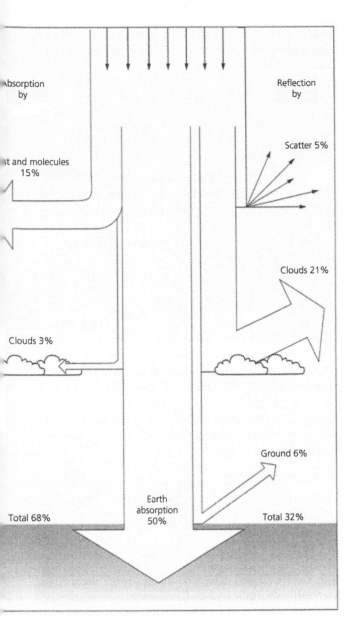

Figure 7.4 *The fate of incoming solar radiation. The figure shows the average effect of various factors that reflect and absorb incoming solar radiation. About half of the energy received at the Earth's surface is eventually released to the atmosphere and re-radiated back into space. After Figure 4.3 in Doerr, A.H. (1990) Fundamentals of physical geography. Wm.C. Brown Publishers, Dubuque*

pollution, especially of particulates, will reduce the incoming solar radiation. Anthropogenic inputs of carbon dioxide will increase the amount of long-wave radiation redirected to the Earth's surface.

2   Figure 7.16 quantifies the global sources of methane, and the major contributing countries, from human activities. Answer the following questions.

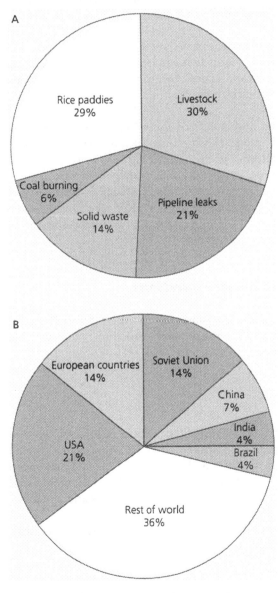

Figure 7.16 *Global sources of methane from human activities. The most important sources are livestock, rice paddies and pipeline leaks (A), and the USA contributes about a fifth of the global total (B). After Box Figure 17.5 in Cunningham, W.P. and B.W. Saigo (1992)* Environmental science: a global concern. *Wm. C. Brown Publishers, Dubuque*

Answers

(a) This may be exemplified by methane production from livestock. However, the principal reason for higher emissions over other countries varies in individual cases. Compared to Europe the USA has a greater natural resource of appropriate land for livestock ranching. Whilst Russia may have a comparable resource its consumer demand does not warrant the same level of livestock rearing. This is true of the rest of the economically developing countries which additionally have less extensive suitable land.

(b) As a hydrocarbon, methane inputs carbon into the atmosphere. It is implicated in global warming, having eleven times more potential as a greenhouse gas than carbon dioxide.

(c) As a by-product of agriculture two strategies may be adopted. Firstly, there might be a change in agriculture systems: reducing cattle ranching and rice production. Secondly, there may be a more applied approach whereby methane is used for local energy production. This is more likely with landfill and intensive rearing systems where the sources are spatially concentrated.

Short-answer questions

1 List the instrumented remote platforms used to monitor the atmosphere and the main variables recorded.

Answer

The main instrumented remote platforms used to monitor the atmosphere are: balloons, aircraft, rockets, satellites and manned spacecraft. They record data on: vertical temperature profiles; changes in trace gases; solar-energy inputs; wind speeds and patterns.

2 Describe why the stratosphere is critically important to life on Earth.

Answer

Stratospheric processes drive the most important global wind circulation systems. These systems distribute atmospheric contaminants around the globe, defining the level of global atmospheric problems. The stratosphere contains the ozone layer which screens the Earth from harmful solar ultra-violet radiation. It is thus critically important for maintaining life on Earth.

3 What is the Earth's albedo?

Answer

Albedo is the ratio of reflected solar radiation from the Earth's surface to that reaching the surface. It is always less than one and is often expressed as a percentage.

4 List the main reasons for expanding the use of solar heating systems.

Answer

The main reason for expanding solar heating systems are: the potential energy supply is inexhaustible; the technology is well developed with future promise of greater efficiency and reduced cost; this system of energy production is less polluting than fossil fuel burning or nuclear energy; and spatially the renewable source is extensively distributed around the world in one usable form or another.

5 What are the main health effects of particulate-based smog?

Answer

High levels of smog have produced large increases in short-term mortality rates in relation to respiratory problems, such as bronchitis and pneumonia and have shown strong associations with heart diseases and strokes.

6 Describe the atmospheric composition changes associated with major air pollution problems.

Answer

Sulphur dioxide from factory chimneys and power stations combines with water in the atmosphere to form sulphuric acid falling as acid precipitation. Inputs of CFCs break down the stratospheric ozone layer, reducing its screening effect. Anthropogenic sources of carbon dioxide and methane increase the proportion of greenhouse gases and enhance the greenhouse effect.

7 What are CFCs and where are they used?

Answer

CFCs are synthetic chemicals mainly comprised of methyl chloroform and carbon tetrachloride. They are inert and essentially non-toxic. Until recently they were used widely as propellants in air-conditioning systems and refrigerators, and in the manufacture of foam boxes and cartons.

8 Describe the main uncertainties surrounding the acid rain problem.

Answer

Firstly, it is difficult to establish the spatial link between point source and deposition area due to the scale and complexity of the atmospheric pathways. Secondly, the actual link between acid input and resultant ecological damage is not fully understood. This is particularly true of critical loads. The importance of this is evident in the policy question: by how much should we reduce acid deposition? Finally, there is incomplete knowledge of the atmospheric chemical reactions, and the conditions instigating them, to be certain that reduced emissions will reduce the acid rain. The last two points emphasize that many environmental relationships are non-linear in associated cause and effect.

9 What is the evidence of global warming?

Answer

Over the last century average world temperatures have risen by 0.5 degrees centigrade. Consequently, global temperatures are now higher than at any time since instrumented records were kept. Recent records show that the four hottest years over the last century occurred recently, in the 1980s.

10 Outline the main problems of implementing reductions in greenhouse gases.

Answer

The scale and sources of emissions are extremely difficult to measure. Factories and power stations can be monitored accurately if instrumentation is installed but area sources such as extensive land use are much more difficult to monitor. Without an inventory of emissions from monitoring it is not possible to see if targets are reached. All this, of course, assumes there is the global political will to agree targets. National sovereignty rights suggest not all countries would agree to target setting, especially those with most to lose economically and socially via development.

Additional references

Hadfield, P. (1997) 'Raining acid on Asia'. *New Scientist* 153 (2069), 16–17.
Good non-Western example of acid precipitation pollution. Discusses the relationship with economic growth and development. Describes the types of anti-pollution equipment available to clean up the source pollutants.

Hagland, W. (1995) 'Solar energy'. *Scientific American* 273 (3), 136–9.
A broad overview of the potential and state of technology required to ensure future use of solar power for electricity generation.

Moulin, C., Lambert, C., Dulac, F. and Dayan, U. (1997) 'Control of atmospheric export of dust from North Africa by the North Atlantic oscillation'. *Nature* 387 (6634), 691–4.
Satellite observations of atmospheric dust transport are correlated through time with variables in the ocean circulation system and large-scale atmospheric circulation system. These mechanisms are seen as influencing other airborne particulate distributions.

Pearce, F. (1996) 'Lure of the rings'. *New Scientist* 152 (2060), 38–40.
Presents the argument for tree ring analysis (dendrochronology) as a major source for the climatic record, especially recent global warming trends. The importance of the greater spatial distribution of trees is stressed, giving better global coverage than other methods, e.g. ice cores.

Pearce, F. (1997) 'Chill winds at the summit'. *New Scientist* 153 (2071), 12–13.
Outlines problems and quantifies targets for the future. Main discussion focuses on the problem of international agreement and the pricing of pollution.

Ramaswamy, V., Schwarzkopf, M. D. and Randel, W. J. (1996) 'Fingerprint of ozone depletion in the spatial temporal pattern of recent lower-stratospheric cooling'. *Nature* 382 (6592), 616–18.

Ozone depletion is seen as the main cause of cooling in the upper atmosphere. This provides another process variable to be accounted for in the complex general circulation model of the atmosphere.

Web site

www.epa.gov/

This US Environmental Protection Agency site offers information on all major air pollutants. It also has information on the status of outcomes such as ozone depletion and global warming.

Aims

- To focus on processes in the lower part of the atmospheric system.

- To signify how these processes redistribute air, heat and moisture across the Earth, producing distinct patterns with and in relation to the other natural systems.

Key-point summary

- Temperature affects environmental processes and systems. Temperature gradients drive the global air and ocean circulation systems. At smaller regional and local scales temperature regimes govern the rates of weathering and erosion.

- Temperature also affects the human physiological system. At this individual scale extremes of heat and cold approach the body's thermal thresholds, producing *heat stress* (at high temperatures) and exposure and *frost-bite* (at very low temperatures).

- Indirectly, temperature affects the operation of various activities in the human economic system. This varies from fundamental components such as agricultural systems – related to plant growth – to extreme temperature regimes promoting tourism based on snow or warm, sunny conditions.

- The patterns of heating relate to spatial diffusion or concentration in relation to incoming solar radiation (*angle of incidence*). The duration of warming by solar radiation also has a temporal aspect controlled by length of day and season.

- The interaction of the Earth's surface and atmospheric system produces four main processes of heating: *radiation, convection, conduction* and *advection*, which display system dynamics and

flows. The different flows and exchanges produced by these processes power the vertical and horizontal movements of air within the atmosphere.

- Global temperature regimes are a reflection of the pattern of heating of the Earth's near-spherical surface (*latitude*). Regionally and locally these are changed by type of surface (sea or land) and *altitude*.

- Changes to the heating/temperature component of the atmospheric system by human-induced warming may produce global system alteration. Regional temperature regimes may shift spatially, disrupting many surface systems.

- The atmosphere is a dynamic system comprising movements of parcels of air. These parcels of air are interlinked and the relationship between adjacent ones governs movement or stasis in the atmosphere. Air will move vertically if there is a temperature difference between it and the surrounding air (*instability*). If the temperature is the same then the air remains stationary relative to its height above the ground surface (*stability*).

- Pressure systems generate horizontal air movements (*winds*). These systems are underpinned by temperature regimes at a variety of scales. High temperatures cause air to rise, thus reducing pressure (*low pressure*), and low temperatures cause air to subside, increasing pressure (*high pressure*). At the global scale latitude governs the distribution of pressure systems. These latitudinal systems have relative pressure changes within them producing short-term *weather* changes and longer-term *seasonal climatic patterns*.

- Wind movement is controlled by the interaction of three global systems. The atmospheric system producing pressure gradients (flows); the Earth's motion system exerting deflection (*Coriolis force*); and the relief system providing *friction*.

- The system of wind movement within the atmosphere is a balance between the pressure gradient (flow from high to low pressure) and the Coriolis force (deflecting this flow). This *strophic* balance produces the general air circulatory system which in turn fuels the Earth's weather systems. The *geostrophic* wind system produced is affected at low altitudes by the friction of relief. Increased friction reduces the effect of Coriolis force, effectively strengthening the flow down the pressure gradient towards low pressure.

- Winds are not confined to the lower atmosphere. In the upper troposphere large-scale circulatory systems both produce winds (*jet streams*) and influence the movement of the lower altitude geostrophic winds via *Rossby waves*.

- As in other systems the global pattern of winds can be spatially defined. Four main belts latitudinally circle the Earth, providing discrete wind conditions within each belt. These belts are dynamic and shift slightly in response to seasonal change.

- The movement of air, both horizontally and vertically, redistributes heat around the Earth's surface. These large-scale thermal flows interconnect and reinforce the global system of wind belts. They also provide general system integrity by preventing overheating at the equator.

- These wind movements link in with other systems. An important example is the movement of water vapour within the global water cycle.

- *Wind energy* has been widely exploited through history. As wind is variable both in time and space around the Earth particular sites have more potential than others for utilizing this energy.

- The *renewable, environmentally clean* nature of wind power means it has relatively little detrimental effect on natural systems. However, its scale of exploitation can affect the environment by using large parcels of land and reducing the aesthetic qualities of an area. The perceived benefits tend to outweigh these costs when considering the larger-scale energy impacts.

- *Atmospheric moisture* is variable through time and space. Its distribution is closely linked to temperature patterns and processes, air movements and wind systems. Warm air has the ability to contain more water vapour than cooler air.

- *Phase changes* convert water from one physical state to another, the most important being *evaporation*, from the surface water system (liquid to gas), and *condensation* around small particles in the atmospheric system (gas to liquid).

- Condensation products vary according to local temperature conditions. They include *dew*, varieties of *frost*, types of *fog* and most obviously *clouds*.

- The outcome of condensation for the Earth's surface is *precipitation*. In the cloud component of the atmospheric system the condensed water vapour requires mixing, usually by *turbulence*, to enable a coalescence of droplets to form a mass big enough to overcome any rising air currents and fall under gravity to the ground.

- Cloud formation and distribution is reliant on the stability of air. *Stable* air does not produce a phase change and hence clouds. *Unstable* air with cooling to the *dew point* produces a condensation phase change.

- The forms of precipitation from clouds cover a wide range of particle sizes and states such as *rain*, *snow* and *hail*. These are mainly controlled by atmospheric temperature conditions.

- Spatial and temporal variations in precipitation quantity may be controlled by the interaction with other natural systems, e.g. *oceanic cold and warm currents*.

- Precipitation quality reflects both natural processes such as volcanic activity and increasingly anthropogenic input in the form of air pollution. These changes in quality affect systems on the Earth's surface.

- Precipitation regimes also result in direct impacts (hazards) to the human system. Short-term events such as severe *hail storms* or *thunderstorms* result in loss of life and economic disruption. Longer-term patterns may produce *drought* conditions with greater social impacts.

Main learning hurdles

Lapse rates

This is an area that usually requires additional explanation to ensure student comprehension. The definitions and distinctions of adiabatic lapse rates, their relationship to water vapour content, and the environmental lapse rate should be discussed initially. Subsequent reference to Figures 8.3, 8.4 and 8.5 should establish a firm knowledge base.

Rossby waves

At this stage, prior to the next chapter on weather systems, it is important that the student has a reasonable understanding of upper-atmosphere Rossby waves. As a minimum they should be aware of their relationship with the jet stream (introduced in the previous chapter) and the general pattern of them in relation to surface pressure. Their importance in the latitudinal transfer of heat should be emphasized.

Key terms

Absolute and relative humidity; adiabatic lapse rates; advection; air currents; Celsius; cirrus; clouds; condensation; conduction; convection; Coriolis force; cumulus; dew point; doldrums; evaporation; Fahrenheit; fog; friction; frost; geostrophic wind; Hadley cells; hail; intertropical convergence zone (ITCZ); isobars; isotherms; jet stream; Kelvin; pressure gradient; radiation; rain; Rossby waves; snow; stratus; temperature; tropopause; the Westerlies; wind; wind energy.

Issues for group discussion

Discuss the distribution of global temperature patterns

An initial resource of a world map of temperature variations should be studied. The discussion should focus on the relative control exerted by latitude, relief and the distribution of oceans and continents. The discussion should develop into considering seasonal change in the distribution. An applied focus could be explored by discussing the differential effects of global warming.

Discuss the best sites for efficient wind power production. What negative effects might there be from using these sites for wind farms?

Initial discussion should examine the distribution of surface wind strengths. The students should also examine other factors in relation to users of the renewable energy source. Chell (1993) is useful in illustrating the wind power resource and providing a means of evaluating the second part of the discussion.

Selected reading

Chell, K. (1993) 'Landscape assessment and wind power'. *Geography Review* 7(2), 23–30.
The potential for developing wind power in Britain is explored in relation to both atmospheric systems and topography. Spatial distributions of both the wind power resource and likely areas of impact are clearly illustrated.

Texts

Barry, R. G. and Chorley, R. J. (1992) *Atmosphere, Weather and Climate*, 6th edn. Routledge: London.
An essential introduction to atmospheric processes and their contribution to the Earth's weather and climatic conditions. Suitable for a range of students and extremely well illustrated to support the clear text explanations.

Kraus, E. B. and Businger, J. A. (1994) *Atmosphere–Ocean Interaction*, 2nd edn. Oxford University Press: Oxford.
A rather advanced scientific text. It provides a useful integration of the processes operating between the two major environmental systems. This scientific system knowledge points to applications relating to managing global environmental problems.

Essay questions

1 Discuss atmospheric instability.
2 Compare and contrast the major characteristics of the troposphere and the stratosphere.
3 Discuss the vertical change in temperature through the atmosphere.
4 Explain the major processes that heat the atmosphere.
5 Give a detailed account of the processes leading to the formation of cloud droplets and explain how these processes may lead to precipitation.
6 Describe and explain the general global wind circulation pattern.
7 Outline the different ways rainfall may be generated.
8 How does the concept of equilibrium help explain atmospheric systems?

9 Evaluate the advantages wind power has over other renewable sources of energy.

10 What are the detrimental environmental effects associated with large-scale wind farms. How may they be reduced?

Multiple-choice questions

Choose the best answer for each of the following questions.

1 Which of the following is an absolute temperature scale?
(a) Centigrade
(b) Kelvin   *
(c) Fahrenheit
(d) Celsius

2 The transfer of energy in the form of electro-magnetic waves is known as:
(a) convection
(b) conduction
(c) advection
(d) radiation   *

3 The normal decrease in temperature with height (lapse rate) is:
(a) 3.5 degrees centigrade per km
(b) 6.5 degrees centigrade per km   *
(c) 9.5 degrees centigrade per km
(d) 16.5 degrees centigrade per km

4 Variations in pressure are depicted on a map by:
(a) isobars   *
(b) isotherms
(c) isotopes
(d) isomers

5 A commonly used name for a horizontal movement of air is:
(a) wind   *
(b) current
(c) wave
(d) stream

6 Which of the following scales measures wind-speed?
(a) Beaumont
(b) Blenheim
(c) Buys-Ballot
(d) Beaufort   *

7 Buys-Ballot defined the relationship between:

(a) windspeed and the ground
(b) Coriolis force and windspeed
(c) pressure gradient and Coriolis force   *
(d) pressure gradient and wind speed

8 Westerlies are generated by:
(a) polar highs
(b) the intertropical convergence zone
(c) subtropical highs   *
(d) equatorial lows

9 High-level, thin, wispy, feather-like clouds are known as:
(a) stratus
(b) cirrus   *
(c) cumulonimbus
(d) cumulus

10 The hail, generated by thunderstorms, that fell on Orlando, Florida, on 25 March 1993 caused damage totalling:
(a) $60 thousand
(b) $6 million
(c) $16 million   *
(d) $60 million

Figure questions

1   Figure 8.12 is a generalized atmospheric circulation model of the northern hemisphere. Answer the following questions.
(a) Explain why atmospheric circulation is confined to the troposphere and what is the prime cause of this circulation?
(b) Account for atmospheric circulation associated with Hadley cells.
(c) What are Rossby waves and how do they influence the development of depressions and anticyclones?

Answers

(a) The troposphere essentially acts as a closed system being overlain in most places by a temperature inversion and in others by the isothermal lower stratosphere. The troposphere contains the majority of the atmosphere's gas molecules (about 70 per cent) as well as water vapour and aerosols from the Earth's surface. These materials allow heat transference, powering the atmospheric heat engine and driving circulation. The prime cause of this circulation is the redistribution of excess energy at

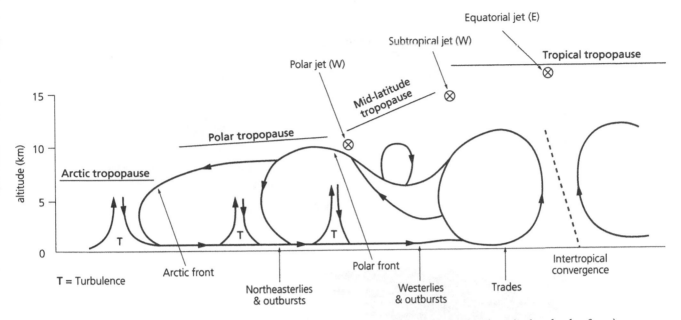

Figure 8.12 *Jet streams and global wind circulation. Three main jet streams (equatorial, subtropical and polar front) are associated with interactions between the tropopause and the major wind systems below. The pattern shown applies to the northern hemisphere. After Figure 14.14 in Dury, G.H. (1981)* Environmental systems. *Heinemann, London*

the equator to the energy deficit at the poles, principally by winds. Heat is dispersed mainly by convection, advection and sinking air.

(b) As the equator is nearest the Sun it is heated more strongly, producing a positive heat budget. This is radiated as long-wave radiation by the surface, heating the surrounding air. The air rises by convection until it reaches the tropopause and flows towards the poles. At about 30 degrees north it cools and sinks. The cell is completed by surface trade winds returning the air towards the equator.

(c) Rossby waves are patterns of air pressure high in the troposphere. Often there may be between three and six identifiable in each hemisphere at any one time. They vary in altitude with an accelerating, diverging upward limb drawing air from the surface and forming a low pressure zone (depression). The downward limb is decelerating and converging, drawing air towards the surface and creating a high pressure zone (anticyclone).

2 Figure 8.16 illustrates common stratiform cloud types. Answer the following questions:
  (a) Outline the three main altitudinal groupings of clouds.
  (b) What are their main characteristics?
  (c) What precipitation conditions are they associated with?

Answers

(a) Clouds may be differentiated by three altitudinal bands within the troposphere. Low clouds form up to about 2,000 m. The middle band of clouds form between 2,500 m and 6,000 m whilst the highest band form above 6,000 m and extend up to about 10,600 m.

(b) Being stratiform cloud types they are well layered. Their rate of upward motion is slow and they usually have a large area extent ranging up to thousands of square kilometres. The low band of clouds is composed of water droplets. Above this, up to about 6,000 ice crystals are mixed in with the water droplets. The highest band of stratiform clouds is composed mainly of ice crystals. Cumulus types of stratiform cloud exhibit weak convection and therefore cover a wider altitudinal range.

(c) The low and middle bands of stratiform cloud type are the main rain-bearing clouds of temperate latitudes. They are associated with the depressional sequence. High-level cirrus clouds signal a weather change accompanying an oncoming depression, the clouds gradually thicken through cirrostratus and altostratus until the nimbostratus bring precipitation. This may be either rain or snow dependent on temperature conditions.

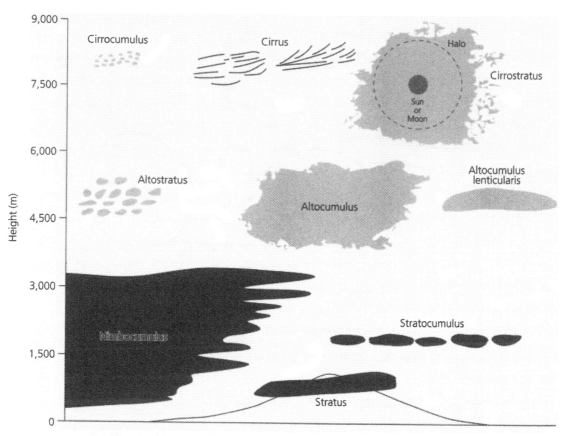

Figure 8.16 *Common stratus cloud types. The figure shows, in schematic form, the main stratiform cloud types. After Figure 5.11 in Briggs, D. and P. Smithson (1985)* Fundamentals of physical geography. *Routledge, London*

Short-answer questions

1 Describe the general pattern of temperature variations.

Answer

Temperature generally decreases from the equator to the poles. At a given latitude temperature will also decrease with altitude. Due to the different rates of heating and cooling of soil/rock and water temperature, ranges tend to be more extreme on land than over the oceans.

2 What is the IPCC 'business as usual' scenario?

Answer

This is a prediction of future climate change made using present patterns as a base-line and extrapolating them into the future. It assumes that there is no change to current human system practices. The Intergovernmental Panel on Climate Change (IPCC) predicts a global mean temperature rise of 0.3 degrees centigrade per decade over the next century. Regionally these increases will not be uniform due to the distribution of land and sea. Land surfaces will warm more rapidly and temperate continental areas are likely to experience the greatest change.

3 Outline the significance of atmospheric stability.

Answer

The relative stability of a parcel of air affects its vertical movement within the lower atmosphere. Therefore, it is important in cloud development and the outcome of precipitation. Apart from this natural atmospheric system operation, air stability influences the speed and extent of air pollutant dispersal and has an influence on the formation of photochemical smog.

4 What is the Coriolis force?

Answer

The Coriolis force is the force generated by the Earth's rotation. It is of importance to the movement of both oceanic water and air in the atmosphere. The resultant force deflects air particles to the left in the southern hemisphere and to the right in the northern hemisphere. Its effect on wind is only directional and does not affect speed.

5 What is a jet stream?

Answer

Jet streams are strong, narrow bands of wind. They occur at high altitude (9,000 m–15,000 m) and are driven by steep thermal gradients. The velocity of most jet streams ranges from 160 to 320 km per hour. Two major jet streams are recognized: the polar jet stream (strong, discontinuous westerly airflow) and the subtropical jet stream (strong, continuous westerly airflow) occurring in latitudes around 30° N and S.

6 What are the Westerlies and why are they important?

Answer

The Westerlies comprise a wind belt located at about 30° latitude north and south of the equator. A combination of poleward pressure gradient and the Coriolis force means their prevailing wind direction is from the south-west (northern hemisphere) and north-west (southern hemisphere). These winds play an important part in moving air masses and associated weather systems from west to east in these mid-latitudes. Weather changes due to the eastward movement of cyclonic cells are the dominant weather pattern associated with the Westerlies.

7 List the main problems associated with wind farms.

Answer

The main problems associated with wind farms are that they may generate noise pollution, affecting local residents; they are often visually discordant and highly visible; and there are perceived health risks due to the generation of electromagnetic interference.

8 Outline the phase changes undergone by water in the atmosphere and identify the processes involved.

Answer

Water in the atmosphere exists in one of three states: as a solid (ice crystals); as a liquid (water droplets); and as a gas (water vapour). Water may change its state from solid to liquid by melting and liquid to gas by evaporation. As a liquid it may become solid by freezing. Gas becomes a liquid by condensation, and a solid by sublimation.

9 Describe the process of condensation.

Answer

The process of condensation is the formation of water droplets around small particles (hydroscopic nuclei) in the air. Condensation is promoted by the availability of such nuclei whose sources may be natural, e.g. pollen, or anthropogenic, e.g. particulate air pollution. Air turbulence also aids the process of condensation by increasing mixing. Collision and coalescence of water droplets increases the individual size until they fall as precipitation.

10 What are the key large-scale characteristics of precipitation distribution?

Answer

Precipitation tends to decrease from low to high latitudes. It also decreases from coastal margins towards continental interiors. In relation to relief the orographic effect tends to promote greater precipitation on the windward slope than the leeward.

Additional references

Hedin, L. O. and Likens, G. E. (1996) 'Atmospheric dust and acid rain'. *Scientific American* 275 (6), 56–60.
An extremely useful account of how atmospheric components and processes influence acid rain distribution. It importantly links geochemical cycles and components within the biosphere to the magnitude and effects of this acid precipitation.

Tsonis, A. A. (1996) 'Widespread increases in low-frequency variability of precipitation over the past century'. *Nature* 382 (6593), 700–2.
Global precipitation patterns are examined decade by decade. Variability of rainfall within these time periods has increased with significant social and economic implications. The general trend is consistent with predicted global climate warming.

Web site

www.badc.rl.ac.uk
This is the site of the British Atmospheric Data Centre. It provides extensive information regarding individual atmospheric variables and contemporary monitoring systems. There are also links to other associated sites.

Aims

- To show that weather conditions are the manifest-ations of processes operating in the atmosphere.

- To show the variety of effects of weather systems and weather events on the Earth's surface systems.

- To explain the factors that produce the spatial and temporal distribution of major weather systems and important individual weather events.

Key-point summary

- Weather is, and has been, a strong influence on society. At a basic level it promotes adaption of the human environment (shelter) to cope with unsuitable conditions.

- The close interaction between weather and people produces a wide variety of responses in the human environment. These include behavioural, eco-nomic, physiological and cultural reactions to particular weather regimes or events.

- A common relationship is with extreme weather conditions (*hazards*) that precipitate *disaster*. The relative uncertainty of the atmospheric system to produce these events either through time or spatially causes severe problems for human system functioning.

- Disruptive weather conditions operate at a variety of scales and may be influenced by atmospheric system changes elsewhere. The enhanced loading of *greenhouse gases* into the atmosphere may shift processes within this system, producing an adaptation in the global pattern of weather. Large-scale *regional* weather changes forcing surface systems to adapt would then ensue, e.g. drought conditions producing soil and vegetation change

(*desert conditions*). Human systems powered by the weather, such as *hydropower* schemes, would also be disrupted.

- The implementation of scientific knowledge of the weather is restricted in time. Short-term *forecasting* and *prediction* have reasonable levels of accuracy, relying on sophisticated process modelling of increasingly more complete monitored data. The functioning of the natural system in a uniform manner is crucial to the accuracy of forecasting.

- *Air masses* are large-scale bodies of air and exhibit general homogeneity at the ground surface/atmosphere interface. Therefore, similar weather conditions are experienced over the spatial area covered by the air mass. Towards the air mass boundary weather changes, as adjacent air masses exert influence.

- Air masses are classified by temperature and humidity characteristics produced from the surface systems underlying them at source. Movements within the atmospheric circulatory system modify these characteristics but they retain enough integrity to influence weather conditions in their path and redistribute heat and moisture crucial to the workings of the global heat system and global water cycle. If an air mass undergoes significant modification due to internal change or external surface contact, the associated weather becomes less predictable. It has changed from a primary to a secondary air mass.

- Large-scale boundaries, between primary, polar and tropical air masses, occur in both hemispheres. These *polar fronts* (their characteristics and position) play a large part in determining weather conditions within the hemisphere during the yearly cycle.

- *Frontal systems* are complex and understanding of

their dynamic workings has involved a large amount of recent scientific data gathering and subsequent modelling. However, general observation of the sequence of weather events associated with fronts has enabled descriptive weather forecasting to be relatively accurate for most of the twentieth century.

- Fronts may extend through the atmosphere to the troposphere (8 km) at a generally gentle slope. Thus, the areal coverage of the weather associated with a front will be extensive, and reliant on the height and slope of the boundary between two air masses.

- The polar front is not latitudinally regular, consisting of a series of tongues related to the motion of *Rossby waves* and the *polar jet stream*. Each lobe will define an area of polar low pressure with the spaces between being occupied by high pressure. Thus *warm* and *cold fronts* are smaller components defined by the polar front system.

- Within this general system, both low pressure cells (*depressions*) and high pressure cells (*anticyclones*) develop. They both spatially define distinct weather patterns and exhibit feedback mechanisms by which they intensify their pressure gradients.

- As with other environmental systems, less frequent intense system conditions produce dynamic effects in other systems. These conditions in the weather system are referred to as *storms*. Associated intense rainfall affects surface systems, magnifying erosion processes and promoting landscape change. Storms are also a hazard to the human system.

- *Thunderstorms* are spatially defined by areas of powerful, convectional uplift related to ground/air heating, active frontal systems and relief. They exhibit temporal patterns at a variety of scales (*seasonal* and *diurnal*), reflecting the processes involved in their formation. As a hazard they are fairly localized and produce heavy rainfall, which may trigger flash floods, and hail. Other hazard links to the surface system are via cloud to ground *lightning*. Lightning performs natural atmospheric system operations, acting as a catalyst for *nitrate* formation. However, alteration of the processes that produce thunderstorms by global warming may increase the lightning hazard by up to 30 per cent.

- As with thunderstorms, *tornadoes* may be spatially and temporally defined by casual atmosphere processes. They have a seasonal and diurnal control and are short-lived, localized, intense

depressions. They are a hazard to the human system due to the generation of very high winds by the local steep pressure gradient.

- *Hurricanes* are larger in scale, both in space and time, and in the main are confined to tropical and subtropical areas. They are intrinsically linked by process to air–sea interactions and by distribution to global wind systems. They link to the human system as a direct atmosphere hazard but also produce secondary hazards (*storm surges*) in coastal areas. Monitoring technology has improved forecasting in susceptible areas and reduced the hazard there by advanced warning and preparation. Extratropical hurricanes produced by rare conditions outside the tropical zone may therefore impact more due to the lack of preparedness for these infrequent events.

Main learning hurdles

Meteorological hazards

The students should review and build on the material from Chapter 8. It should be clear that atmosphere processes underpin the weather systems described. The variations in the intensities of the processes underpin the hazards outlined at the end of the chapter. It is useful to tie in the initial contribution of the solar-energy system and the surface systems, though this should be largely evident from the previous chapter.

Key terms

Ana-fronts; anticyclones; cold front; cyclogenesis; depressions; fronts; hurricanes; isobars; kata-fronts; lightning; occlusion; polar/Arctic air masses; polar jet stream; storm environment; storm surges; synoptic; thunderstorms; tornadoes; tropical air masses; warm front; weather forecasting.

Issues for group discussion

Discuss how the weather influences you

Initially the students should focus on their own experience, usually of being hot or cold, wet or dry. Reading Kalkstein and Davis (1989) might lead them to appraise the question over longer time-scales. Further emphasis must focus on broadening out the indirect effects, e.g. food production and cost, as well as the potential infrequent hazardous event. Reference to Burt and Coones (1990) is useful for this second part.

Discuss the effects of hurricanes/cyclones on different groups of people

The students should be encouraged to have a base knowledge of the likely impacts, e.g. Smithson (1993). From this they may discuss hurricane effects in countries with different economic development status. Two 'geographies' may be usefully emphasized. The first is the single event that affects both types of country, e.g. a Caribbean and United States case study. Secondly, the overall hazard problem between two countries, e.g. Bangladesh and Japan case studies. Reference to Bridges (1988) should emphasize the impact on less economically developed areas.

Selected readings

Bridges, E. M. (1988) 'Rarotonga after Cyclone Sally'. *Geography* 73 (2), 154–7.
A case study of the effects of a severe weather event on natural systems. Also outlines the spatial damage in relation to the storm path and topographical influences.

Burt, T. and Coones, P. (1990) 'Winds of change? The gale of 25 January 1990'. *Geography Review* 3 (5), 22–9.
A comprehensive article looking at more frequent severe wind hazards as evidence for climatic change. Very useful spatial analysis with a discussion of statistical probability and event forecasting.

Kalkstein, L. S. and Davis, R. E. (1989) 'Weather and human mortality: An evaluation of demographic and interregional responses in the United States'. *Annals of the Association of American Geographers* 79 (1), 44–6.
An interesting account of general weather variables, especially temperature, correlating hazards to human communities. This produces an important counterpoint to the more extreme weather conditions usually associated as hazardous. Spatial overlap between human system components and weather systems are demonstrated.

Perry, A. and Reynolds, D. (1993) 'Tornadoes: The

most violent of atmospheric phenomena'. *Geography* 78 (2), 174–8.
Illustrates the lack of knowledge relating to smaller-scale violent weather processes within the global climatic system. Brings together observational event change with likely effects for climate change. Clear classification of the scale of damage related to actual wind speed is described.

Smithson, P. (1993) 'Tropical cyclones and their changing impact'. *Geography* 78 (2), 170–4.
The distribution of these severe weather systems is reviewed in space and time. Human loss is equated through time, and hazard prediction is discussed.

Textbooks

This area is covered by the atmospheric processes texts (Chapter 8) and the climatological texts (Chapter 10).

Essay questions

1 For mid-latitude temperate areas, discuss the weather patterns that produce hot dry summers and warm wet winters.

2 Discuss the types of rain storm that produce high rainfall intensities and account for their spatial and temporal extent.

3 Contrast warm and cold fronts with reference to their relationship to adjacent air masses of different temperatures.

4 Discuss in detail the structure of a North Atlantic depression and outline the weather sequence associated with its passage over a fixed point on the British coast.

5 Evaluate the influence of the weather on *two* human economic activities.

6 Describe the global pattern of hurricane distribution. What areas might face an increased hurricane hazard and why?

7 Modern technology has increased the accuracy of hurricane forecasting and monitoring yet worldwide death tolls have not reduced. Why?

8 Describe the formation of tornadoes and assess them as a spatial hazard.

9 Evaluate the mechanisms by which tropical cyclones may damage a coastal area.

10 Why is the medium- to long-term prediction of intense tropical storms so difficult?

Multiple-choice questions

Choose the best answer for each of the following questions.

1 Bjerknes in 1917 proposed the basic model that describes and explains:
  (a) the role of Rossby waves
  (b) jet stream distribution
  (c) tornado formation
  (d) weather changes at fronts   *

2 Stationary fronts often bring long spells of:
  (a) rain
  (b) unsettled weather   *
  (c) sunshine
  (d) snow

3 Southern England experiences most thunderstorm activity in:
  (a) August
  (b) November
  (c) June   *
  (d) May

4 In a typical year, this decade, the number of cloud to ground lightning flashes observed and monitored by equipment in the USA was:
  (a) 100 thousand
  (b) 1 million
  (c) 5 million
  (d) 15 million   *

5 Lightning plays an important role in:
  (a) the nitrogen cycle   *
  (b) the carbon cycle
  (c) the sulphur cycle
  (d) the hydrological cycle

6 The most intense hurricane recorded during the twentieth century was:
  (a) Hugo
  (b) Gilbert   *
  (c) Sally
  (d) Andrew

7 Divergence of air at the Earth's surface causes:
  (a) clouds
  (b) anticyclones
  (c) fronts
  (d) depressions   *

8 In the North Atlantic a depression has winds that are:

  (a) diverging and flow clockwise
  (b) converging and flow anticlockwise   *
  (c) diverging and flow anticlockwise
  (d) converging and flow clockwise

9 The most frequent hazardous storms are:
  (a) tornadoes
  (b) extra-tropical hurricanes
  (c) tropical cyclones
  (d) thunderstorms   *

10 In the Indian Ocean a tropical cyclone is called a:
  (a) cyclone   *
  (b) hilly-willy
  (c) hurricane
  (d) typhoon

Figure questions

1 Figure 9.7 shows the distribution of tornadoes in the USA, 1953–1976. Answer the following questions.
  (a) Describe the general distribution pattern.
  (b) What air masses influence this distribution?
  (c) Briefly describe the associated weather, spatial extent and life span of a tornado.

Answers

(a) Tornadoes appear most common in the central to South Midwest and along the East Coast. The number of tornadoes is much fewer in the West.

(b) Cold continental polar air meeting maritime tropical (gulf) air accounts for the high rate of tornadoes in Texas and the Midwest. Maritime tropical (Atlantic) warm air meeting the cold continental polar air accounts for tornadoes on the southern East Coast with maritime polar (Atlantic) cold air meeting maritime tropical warm air to produce tornadoes in the north East Coast states.

(c) Tornadoes generally form when warm, humid air is sucked into a low-pressure cell. A steep pressure gradient is created with a very low pressure centre. This continues to suck in surrounding air and very high wind speeds are generated. These conditions are associated with cold front thunderstorms and tornadoes originate on their periphery. Observations suggest formation at the base of cumulonimbus clouds or within towering cumulus clouds. The

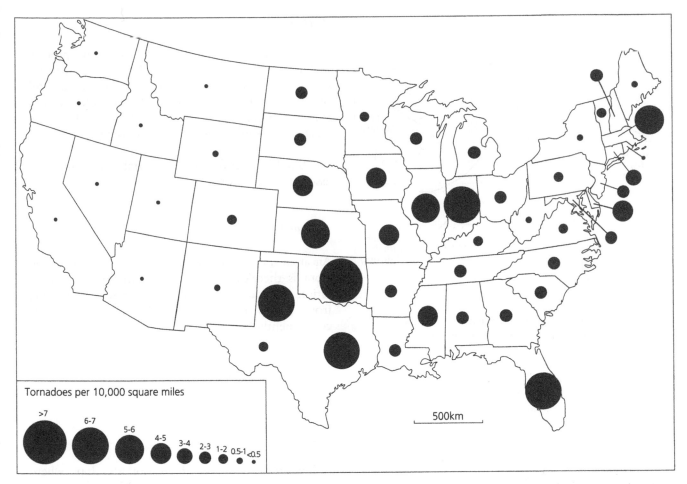

Figure 9.7 *Distribution of tornadoes in the USA, 1953–1976. Tornadoes are most common in the Midwest and along the eastern coast. After Figure 6.12 in Marsh, W.M. (1987)* Earthscape. *John Wiley & Sons, New York*

vortex of a tornado is usually only a couple of hundred metres in diameter, moving in relatively straight paths and lasting only a few hours.

2 Figure 9.1 illustrates the main air masses that affect North America. Answer the following questions.

(a) Explain what happens to the stability of the following air masses during the season shown as they move towards the continental interior:

i Maritime tropical (Atlantic) in winter
ii Continental tropical in summer.

(b) Why is the frequency of anticyclones greater in the northern Midwest states than the west coast of Canada?

Answers

(a) i Maritime tropical (Atlantic) air is moving

from warmer to cooler latitudes and is thus cooling from below, giving increased stability. In winter it encounters the polar front and rises over the cooler polar air. Thus cyclogenesis results in instability.

ii This air is hot and dry at its source area in Mexico, moving northwards into cooler latitudes. It will cool slightly from below but continentality will prevent a significant rise in humidity and it would remain stable.

(b) The northern Midwest has a more southerly location than the west coast of Canada; therefore, the frequency of depressions associated with the polar front is likely to be less. As a consequence, the frequency of blocking anticyclones is likely to be higher. The northern Midwest has a continental location and is more likely to be affected by continental high-pressure systems.

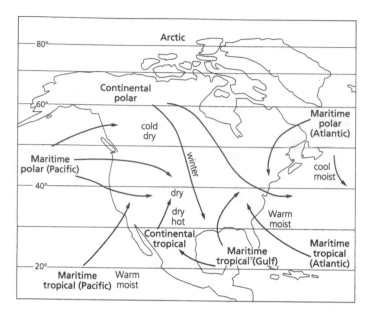

Figure 9.1 *Source regions of the main air masses that affect North America. Each source region generates air masses with distinctive properties (particularly temperature and humidity), and as these move across a region they can bring change in weather. After Figure 4.28 in Doerr, A. H. (1990)* Fundamentals of physical geography. *Wm.C. Brown Purlishers, Dubuque*

Short-answer questions

1 How is climate defined?

Answer

Climate is defined as the long-term prevailing weather conditions of an area. These reflect spatial parameters expressing global and surface systems such as latitude, degree of continentality and altitude.

2 What are the key properties of air mass source regions?

Answer

Air mass source regions have two important features. Firstly, calm conditions allow the air mass to develop relatively uniform properties over time. Secondly, a large spatial extent of similar surface conditions, such as a desert, allows the air mass to build to an influential extent.

3 Outline the causes of change in air masses.

Answer

An air mass may be changed by both internal process and external system variables. Internally, adiabatic change or subsidence associated with instability will modify the air mass. Externally, variable ground surface conditions such as relief or land to sea change will modify the air mass by contact.

4 Briefly describe the two stability conditions associated with fronts.

Answer

Air at an unstable front (ana-front) rises rapidly. The front is therefore very active, with uplift, condensation and identifiable weather changes. At a stable front (kata-front) the air sinks. Condensation is consequently limited and changes in the weather are suppressed.

5 Outline the basic formation of a low-pressure cell.

Answer

A low-pressure cell or depression develops where two air masses converge. In mid-latitudes this is usually at the polar front. Here, small unstable irregularities in the front cause a localized drop in air pressure. The irregularities grow as warm air to the south moves north and rises as the cold air slides underneath it. The air movements are generated by the initial movement of the flowing warm air and the subsequent cold air flowing in behind it. These develop a rotating air mass with the lowest pressure at the centre.

6 What is an anticyclone?

Answer

An anticyclone is a body of air of higher pressure than the surrounding air. The pressure decreases away from the centre. They are formed by the convergence of air in the upper layers of the atmosphere and the associated subsidence and divergence of this air at lower levels. Anticyclones reflect conditions of stability with calm and settled weather.

7 Describe the lightning hazard.

Answer

Lightning is associated with thunderstorms and becomes a hazard when it discharges to the ground. Thunderstorm frequency therefore governs the natural risk but human activities increase the risk from the hazard. Most of the increased risk is from either working or undertaking recreation outdoors. Apart from direct strikes, lightning instigates natural fires (secondary hazard) destroying vegetation and threatening lives. Fires may also be started in inappropriately protected buildings built of or containing flammable materials.

8 List the common names for severe storms in different parts of the world.

Answer

Severe storms are known as: hurricanes in the western hemisphere, typhoons in much of Asia and across the Pacific, cyclones in the Indian Ocean and Australia and willy-willies in some South Pacific islands.

9 Describe what happened during Hurricane Gilbert, September 1988.

Answer

Hurricane Gilbert formed in the eastern Caribbean Sea as a tropical storm. It moved west and badly affected Jamaica as a Force 3 storm (185 kmph). Its main landfall was on the Yucatan peninsula two days later, by which time it had intensified to the lowest pressure (885 mb) on record for a hurricane in the western hemisphere. Associated wind speeds were 280 kmph (Force 5). Here it destroyed tourist resort infrastructure and devastated marine ecosystems and coastal zone habitats with an accompanying storm surge. Inland power lines, farming land and tropical rainforest were damaged. Within 48 hours it had crossed the Gulf of Mexico and hit the town of Monterrey with 190 kmph winds killing about 200 of the total of 318 people killed by the hurricane. It

died down the following day, having caused billions of dollars of damage.

10 Why are Arctic hurricanes a less severe hazard than tropical hurricanes?

Answer

Arctic hurricanes have less energy than tropical hurricanes. Their vertical development is therefore much less and air pressure towards the centre is higher, reducing wind speeds to about 70 per cent of those in tropical hurricanes. Spatially they are smaller and therefore affect a reduced area. The areas affected are also less densely populated and also less developed. Arctic hurricanes last a comparatively short time, less than a day, reducing the temporal risk.

Additional references

Davies-Jones, R. (1995) 'Tornadoes'. *Scientific American* 273 (2), 34–40.
Outlines the present knowledge of the atmospheric processes that produce tornadoes. These are linked into larger-scale atmospheric components and the ground/air interface of the hazard is discussed.

Slattery, M. and Burt, T. (1997) 'The 1995 Atlantic hurricane season'. *Geography Review* 10 (3), 12–17.
The second most active Atlantic hurricane season in recorded history is analysed. Excellently illustrated with a clear analysis of the processes of hurricane formation.

Web site

www.nottingham.ac.uk/pub/sat-images/meteosat
This site provides excellent images and video clips of weather systems recorded by Meteosat. Individual events such as hurricanes are portrayed with complementary data-sets. Provides easy links to other meteorological sites.

Aims

- To relate climate to weather systems and atmospheric processes by comparison of the time-scales of conditions prevalent over space.

- To examine climate regimes over long time-scales and over broad spatial areas.

- To explore the factors associated with climate variability at different scales.

- To examine the evidence for reconstructing the past climate record.

Key-point summary

- Climate is a product of the four major environmental systems (the atmosphere, lithosphere, hydrosphere and biosphere). The interactions between these systems to produce climate and the impact back of climate on these systems are core aspects of the *Gaia* hypothesis. The climatic system may, therefore, be viewed as dynamic and any inputs to the system will be relayed elsewhere.

- Climate is a truly complex global system requiring sophisticated scientific modelling to predict the likely outcomes (changes) to any disturbance.

- People and climate have a symbiotic relationship. Human activities, such as land-use changes and air pollution, alter the natural atmospheric processes that help regulate climate. Climate impacts directly on people in beneficial ways, e.g. agricultural opportunity, and detrimental ways, e.g. climatic hazards.

- Climate exhibits a spatial pattern at a range of scales. The global pattern is established by large-scale system influence, i.e. solar-energy input governed by the Earth's movements in relation to the Sun and the distribution of oceans and continents linked to the plate movement system.

- Superimposed on this general pattern are local and regional variations related to different processes operating within and between the Earth's major environmental systems:

- ocean currents of varying temperatures driven by wind and pressure systems;

- climatic oscillations producing recurrent but short-term variations in regional climate, e.g. El Niño. Changes in one part of the system may effect changes elsewhere in the climatic system (*teleconnections*);

- altitude affects climate locally by exerting influence on temperature, precipitation and winds. Large mountain belts will exert this influence over a regional area.

- The global wind systems have a strong influence on climate by moving heat and moisture through the atmospheric system. These are modified by local/regional conditions to produce smaller-scale temporal patterns ranging in duration from seasonal (*monsoons*) to diurnal (*land/sea breezes*).

- Threatening the intricate workings producing this mosaic of climate are human actions. Initially localized air pollution inputs were generally confined to local atmospheric space. Increasing numbers of sources and larger-scale inputs have started to affect global climate. Land-use change disturbs the biosphere system at the local/regional scale; though the increasing degree of change both through time and space, e.g. tropical deforestation, has global climatic implications.

- At a smaller scale humans may produce a completely artificial change such as urbanization. With this change natural equilibrium is completely broken down with an alteration in system flows of both the local water cycle and atmospheric system.

- Climatic regions may be spatially defined. A global classification will necessarily have to generalize a complex reality into a simple, usable model. *Köppen*'s classification reflects what might be considered the two most fundamental variables affecting the biosphere – temperature and precipitation.

- Climate classification schemes seek to define sub-system boundaries within the general climatic system. These are usually defined on the basis of threshold values, e.g. *temperature*. Whilst not accounting for the reality of gradual change between climatic zones it has the benefit of understandably portraying a very complex global system.

- *Tropical climates* are subdivided by their amount and variability of precipitation. Labelled climatic processes within the tropical zone, i.e. monsoons, illustrate clear relationships to the global atmospheric system, e.g. pressure systems, and to the land–sea system.

- The climates of *mid-latitudes* are characterized by variability.

- *Subpolar and Arctic climates* are divided in relation to their surface characteristics, reflecting continentality and marine surface conditions (open water areas and permanent ice pack). The differentiation between Arctic and subpolar is reinforced by latitude and the permanence of land ice, e.g. ice caps.

- Our knowledge of long-term climate change is based on *environmental construction*. Generally, evidence of past climates is more comprehensive and reliable as we move closer to the present. A wide range of evidence is available and it may be classified under *direct observation*, *historical records* and *proxy indicators*. The nature and quantity of these will vary spatially as well as temporally. Physical and biological proxy indicators clearly indicate the close global system link between climatic condition and surface systems.

- Patterns of climatic change reflect major environmental system interactions and provide baseline cycles at a variety of timescales in which to fix future change. Long cyclical patterns are linked to the ocean–land distribution system driven by plate tectonics. This affects the surface *glacial sub-system*. Within these large time-scale cycles smaller, variable patterns are superimposed, e.g. the Little Ice Age, with evidence supplied by surface and atmospheric depositional material (*sediments* and *ice cores*).

- Causes of climatic change are both internal and external to the climate system. The internal composition of gases and particulates may change with effects on climate. Externally the Sun and cosmic system have global influence on energy input, driving the atmospheric processes that define climate. Whatever the initial cause, a key factor of climate change is that, once instigated, feedback mechanisms reinforce this change. At present the major concern is that relatively recently the natural processes causing climate change and maintaining a Gaian self-regulation have been compromised by human activities affecting the internal composition of the atmosphere. Modification is clearly evident at the regional scale, if yet to be conclusively proved at the global scale.

Main learning hurdles

Environmental determinism

There is a danger with the topic of climate that the concept of environmental determinism becomes too rigidly embedded in the student's mind as to how natural–human interactions take place. As *The Environment* illustrates, the Earth is more complex than this. The instructor should therefore emphasize the latter part of the chapter to stress the anthropogenic influence.

Climate controls and spatial representation

The instructor should make clear what depiction of reality the students are viewing in relation to actual climatic reality. Too often students de-emphasize the zonal characteristics by over-concentration on the fixing of boundaries. The instructor must reinforce the gradual transition between zones and their spatial variability. Some students tend to concentrate simplistically on the latitudinal climate controls, probably gleaned from the seasonal cycle, at the expense of the continentality control. However, the resources used to explain continentality must be chosen with care. The use of a globe is really necessary (though a variety of satellite images is a useful surrogate) to ensure that land masses are depicted in a spatially correct manner and not as distortions of a map projection.

Key terms

Aerosols; altitude; boundaries; chinook; climate zones; Climatic Optimum; dendroclimatology;

determinism; Flandrian Interglacial; Föhn; geological evidence; glacials; Gulf Stream; Holocene; ice cores; interglacials; isotope analysis; Köppen; land and sea breezes; mistral; monsoon; ocean currents; periglacial; pluvial; pollen evidence; symbiotic; teleconnections; tropical climates.

Issues for group discussion

Discuss the range of human interference affecting local climates

The students must consider this widely. Initial focus will be at the local scale, probably concentrating on urban climates. This should broaden out to consider wider-scale land-use affecting soil and vegetational cover. Emphasis towards the end of discussion must focus on how local climate is interdependent with larger global systems upon which humans have an effect, e.g. trans-boundary air pollution.

Discuss the differences and similarities between climate classification schemes

The students should be shown illustrations of the Thornthwaite and Strahler schemes to view alongside the Köppen scheme (Figure 10.6). The discussion should take two lines of enquiry. Firstly an examination of the variables used and the discussions of classification produced. Secondly, the ease of application of the schemes and the identification of commonly recognized and depicted zones.

Discuss the evidence for short-term climate change

It would be useful for the students to read Thompson (1989) to gain an overview and then to look at the case study of Gregory (1993). Students should emphasize the spatial nature of the evidence and why it may or may not be applicable at the global scale. A wider discussion of anthropogenic influences and their management might be developed if time allows.

Selected reading

Gregory, S. (1993) 'A hundred years of temperature and precipitation fluctuations at Sheffield, 1891–1990'. *Geography* 78 (3), 241–9.

A practical paper discussing how local long-term climate records may be used to establish evidence for wider climatic change. Well illustrated with data tables, graphs and simple statistical tests.

Terjung, W. H. (1976) 'Climatology for geographers'. *Annals of the Association of American Geographers* 66 (2), 199–222.
An overview of the topic and its applications.

Thompson, R. D. (1989) 'Short-term climatic change: Evidence, causes, environmental consequences and strategies for action'. *Progress in Physical Geography* 13 (3), 315–47.
A holistic appraisal of climatic system change, its impact on other systems and how these may be managed to alleviate impacts. Current theories of climatic change are discussed with particular links to the coastal zone, agricultural systems and human health.

Textbooks

Barry, R. G. (1992) *Mountain Weather and Climate*. Routledge: London.
A comprehensive text of an often under-represented aspect of climate – relief. Well illustrated with a variety of examples from various global mountain environments.

Lamb, H. H. (1995) *Climate, History and the Modern World*, 2nd edn. Routledge: London.
Discusses the current state of scientific knowledge of climate past and present. Contains good sections on the relationships between atmospheric processes and climate, environmental reconstruction and global warming.

Roberts, N. (1989) *The Holocene: An Environmental History*. Blackwell: Oxford.
This text provides a detailed review of the natural environmental change during the Holocene. It effectively conveys the range of evidence used for environmental reconstruction.

Thompson, R. and Perry, A. (eds) (1997) *Applied Climatology: Principles and Practice*. Routledge: London.
An up-to-date series of specialist topics focusing on the application of scientific knowledge to the study of human impacts on climate. Interaction between the major environmental systems is discussed with an emphasis on climate change.

Essay questions

1 Account for the distribution of the Earth's main climatic zones.

2 Discuss the perceived role of the oceans in controlling global climate.

3 Critically evaluate the application of the orbiting theory in explaining long-term climatic change.

4 Discuss the role of ice cores as sources of information on past climatic changes.

5 Evaluate the main controls on the climate of the North American continent.

6 Examine the critical role of feedback processes in determining global climatic response to external influences.

7 Evaluate the main evidence available for reconstructing the history of industrial atmospheric pollution.

8 To what extent does the increase in scientific knowledge of long-term climatic variation improve the reliability of predicting climate change in the next century?

9 Evaluate the main interactions between climate and other environmental systems.

10 How might climate affect human health?

Multiple-choice questions

Choose the best answer for each of the following questions.

1 A katabatic wind that descends from the former Yugoslavia into the Adriatic Sea is called:
   (a) the bora   *
   (b) the taku
   (c) the mistral
   (d) the chinook

2 Köppen's climate classification scheme is based primarily on:
   (a) temperature and humidity data
   (b) precipitation and wind data
   (c) pressure and sunshine data
   (d) temperature and precipitation data   *

3 Which of the following has not produced a climate classification scheme:
   (a) Strahler
   (b) Dudley Stamp   *
   (c) Geiger
   (d) Thornthwaite

4 How many major climatic groups are identified by the Köppen scheme:
   (a) 15
   (b) 9
   (c) 6   *
   (d) 4

5 The most recent epoch is called the:
   (a) Quaternary
   (b) Pleistocene
   (c) Medieval
   (d) Holocene   *

6 Rainfall variability is greatest in:
   (a) semi-arid areas   *
   (b) equatorial regions
   (c) temperate regions
   (d) Antarctica

7 Which of the following experiences the largest annual temperature range:
   (a) the Asian continental interior   *
   (b) the equator
   (c) the North Pole
   (d) the Pacific Ocean

8 Tropical arid climates are classified as a:
   (a) microclimate
   (b) genoclimate
   (c) macroclimate   *
   (d) mesoclimate

9 The El Niño phenomenon occurs in the:
   (a) Indian Ocean
   (b) Arctic Ocean
   (c) Pacific Ocean   *
   (d) Atlantic Ocean

10 Climatic change during the Pleistocene Ice Age is reconstructed with evidence from:
   (a) oxygen isotopes in marine sediments   *
   (b) historical records
   (c) peat cores
   (d) meteorological records

Figure questions

1 Figures 10.2 and 10.3 illustrate small-scale diurnal wind systems. Answer the following questions.
   (a) Compare and contrast the development of sea breezes and valley breezes during the day.
   (b) Compare and contrast the development of land breezes and mountain breezes at night.

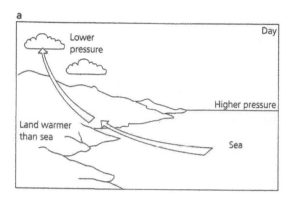

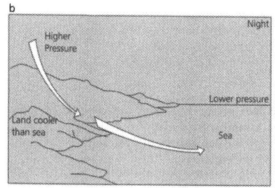

Figure 10.2 Land and sea breezes. See text for explanation. After Figure 6.6 in Doerr, A.H. (1990) Fundamentals of physical geography. *Wm.C. Brown Publishers, Dubuque*

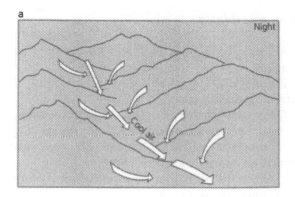

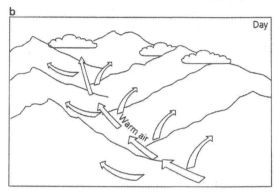

Figure 10.3 Mountain and valley breezes. See text for explanation. After Figure 6.7 in Doerr, A.H. (1990) Fundamentals of physical geography. *Wm.C. Brown Publishers, Dubuque*

Answers

(a) Both sea and valley breezes reflect differential heating and cooling with the resultant pressure gradient producing localized winds. However, there are differences in the parts of the surface system producing these winds and also variations in local weather effects. The high pressure area instigating the sea breeze is due to the slower warming of the sea compared to the land during the day. Thus it represents the differential heating of different surface material. Valley breezes are generated by differential heating related to altitude, slope aspect and vegetation cover. Altitude is the principle mechanism, with the other two controlling the scale of the effect such as a north-facing valley slope heating less and producing a less steep pressure gradient. Both breezes produce unstable air conditions with the less steeply rising sea breezes producing advection fogs whilst strong updraughts in mountain areas may produce cumulo-type clouds with thunderstorms under warm conditions.

(b) At night the process is essentially reversed with a land breeze being generated by the faster cooling land surface. The mountain breeze again reflects an altitudinal control with the cooling air flowing down-valley under gravity. This produces a temperature inversion which under moist conditions may create fog.

2 Figure 10.8 shows the air flows during the summer monsoon over the Indian subcontinent. Answer the following questions:
   (a) In general terms describe the factors producing the conditions shown in the diagram.
   (b) What would be the air flow pattern in winter and why?

Answers

(a) The increase in insolation as the overhead position of the Sun moves northward in the summer creates a low-pressure area over northern India. Warm moist air from the Indian Ocean is drawn towards this area. This humid, unstable air is conducive to precipitation. Over the Western Ghats this is instigated by

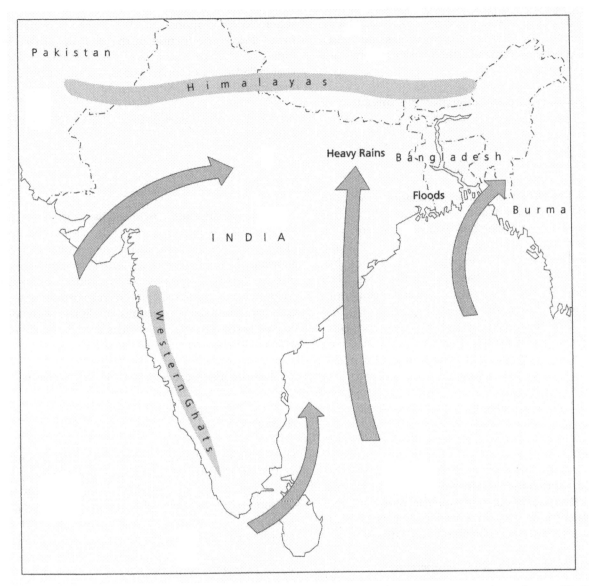

Figure 10.8   Summer monsoon air flows over the Indian subcontinent. Warming air rises over the plains of central India in the summer, creating a low pressure cell that draws in warm, wet oceanic air. As this moist air rises over the Western Ghats or the Himalayas, it cools and heavy rain falls. After Figure 17.9 in Cunningham, W.P. and B.W. Saigo (1992) Environmental science: a global concern. *Wm.C. Brown Publishers, Dubuque*

orographic uplift, whereas in the north-east area both orographic (Himalayas) and convectional uplift (due to the low-pressure zone) produce heavy rainfall.

(b) In winter the conditions are essentially reversed. With the overhead Sun migrating towards the Tropic of Capricorn, the ITCZ low-pressure zone is located in the Indian Ocean. At the same time central Asia is experiencing intense cooling (winter continentality), producing high pressure. The source area is dry and subsequently the south-easterly flowing winds produce little rainfall over the Indian subcontinent.

Short-answer questions

1 What climatic effects on the hydrosphere affect people?

Answer

Climate affects the distribution of ocean currents and changes sea level. This may affect people by influencing coastal weather systems and changing the distribution and production of fish resources.

Both reflect ocean current location. Sea-level change will influence the effect of storms surges and coastal erosion processes, both of which have economic and life-threatening cost.

2 How does climate influence the migration of people?

Answer

Climate has effects on human migration over both long and short time-scales. Permanent migration is encouraged by a pleasant warmer climate with associated reduced economic costs, e.g. heating. Both individuals and companies will, if possible, relocate, such as to the 'sunbelt' in the USA. Over the short term particular climatic conditions encourage a range of tourist migration to cold climates promoting skiing and to warmer climates catering for 'sun worship'. These are often perceived as seasonal activities but many may be viable throughout most of the year.

3 What are teleconnections?

Answer

These are simultaneous atmospheric events in areas remote from each other. From the observation of these, positive or negative correlations may be made with atmospheric/oceanic circulation patterns.

4 How does altitude affect temperature?

Answer

As the atmosphere is mainly heated by ground radiation, air temperature decreases with altitude at the normal lapse rate of 6 to 6.5 degrees centigrade per 1,000 metres.

5 What are the main microclimate changes experienced in urban areas relative to adjacent countryside?

Answer

Urban areas are generally warmer on calm, clear nights due to the heat being radiated from buildings. Relative humidity is lower due to the lack of surface water (lower evaporation). Wind speeds overall are lower due to the friction of the built-up surface but high-rise 'urban canyons' may produce local, channelled gusts of higher speed. Urban air pollution may contribute to higher precipitation potential.

6 Outline the benefits of a climate classification scheme.

Answer

A climate classification scheme allows us to rationalize an extremely complex system. It allows the total global space to be subdivided into regional climates, allowing an appraisal of how climates change between places. Spatial relations with other important environmental systems can be recognized and an examination of the interactions of climate with them may be perceived. It also provides a useful baseline by which to visualize the rate of climate change.

7 List the historical evidence of climate change.

Answer

The historical evidence relates to a documentary record of climate observations. These might be: direct recording of variables from instrumental measurement; weather observation diaries; travel accounts; agricultural records; and the general recording of infrequent phenomena.

8 List the geological evidence of climate change.

Answer

Evidence from the geological record includes: glacial deposits (tills and moraines) indicating the presence of moving ice; palaeosols; former dune patterns outside present sand desert areas; relic lake shorelines indicating wetter conditions; ice wedges and solifluction lobes indicating former periglacial conditions; and the ocean sediment record.

9 What was the Devension (Wisconsin)?

Answer

This was the most recent glaciation. It began about 70,000 years ago and ended about 10,000 years ago.

The glacial maximum occurred about 18,000 years ago where, compared to present climatic conditions, over most of North America and Eurasia south of the ice sheets temperatures were lowered by 8–15 degrees centigrade and in adjacent seas by 2–2.5 degrees centigrade.

10 What are the likely consequences of global warming?

Answer

Likely consequences of global warming could include latitudinal displacement of vegetation belts polewards. Mountain vegetation zones might similarly migrate higher. Tropical storms affected by ocean–atmosphere interaction may increase and affect other areas. All elements of the hydrological cycle would change, with the potentially most serious consequence being the release of water from the global ice store.

Additional references

Pearce, F. (1996) 'Deserts on our doorstep'. *New Scientist* 151 (2037), 12–13.
A spatial outline of the contemporary effects attributed to climate change in the Mediterranean. Potential future problems are also suggested from the present evidence.

Rahmstorf, S. (1997) 'Ice-cold in Paris'. *New Scientist* 153 (2068), 26–30.
Uses a systems approach to explain the importance of the oceanic current system in controlling our climate. A clear explanation of feedback mechanisms in relation to these systems suggests a breakdown of the climatic system with a new lower temperature equilibrium the norm for Western Europe.

Ramstein, G., Fluteau, F., Besse, J. and Joussaume, S. (1997) 'Effect of orogeny, plate motion and land–sea distribution on Eurasian climate change over the past 30 million years'. *Nature* 386 (6627), 788–95.
A well-illustrated article that examines large-scale system interaction over long time periods. The interaction of the lithosphere, hydrosphere and atmosphere are simulated using a general circulation model. This is compared with various palaeoclimatic records.

Web site

www.doc.mmu.ac.uk/aric/
An excellent site that contains information on global climatic change in 'fact sheet' form. These cover all topical sections in this chapter of *The Environment*.

Aims

- To describe and explain the general system workings of the water cycle.

- To assess the role of river systems within this cycle.

- To examine river responses through time to system changes.

Key-point summary

- Water is essential to all forms of life. Knowledge of the functioning of the hydrological cycle is therefore vitally important.

- Within the hydrological system water is stored in oceans, lakes and clouds, and flows, e.g. rivers and rainfall. This system operation can be viewed at two fundamental scales – the global cycle and the *drainage basin* cycle.

- The functioning of the global hydrological system integrates components of the other major environmental systems and influences processes within them. It is driven by atmospheric processes (precipitation and wind) and the surface process of evaporation. With the interaction of all the major environmental systems as components, this continuous recycling of water is extremely complex, containing many variables. Responses within the system to human impacts are therefore difficult to predict completely.

- Relatively small amounts of water are recycled quickly with most being retained in long-term *stores* such as the oceans, ice caps and rocks. This promotes short-term (yearly) system stability at the global scale. Fluctuations within the yearly cycle will occur due to variations in regional weather changes.

- Human activities can influence all stages of the water cycle. Air pollution can increase cloud formation by providing condensation nuclei. Deliberate modification may be attempted by *cloud seeding*. Evaporation rates may be increased by *urban development* or increasing surface water area (*reservoirs*). Changes to river networks and catchment area will affect runoff, e.g. *river regulation* and *land-use change*. Pumping and use of groundwater will lower the water table, reducing the natural storage time. These effects on natural stores and flows still have uncertain outcomes.

- *Drainage basins* are fundamental spatial units. Having a clear, physical boundary means that their surface area can be easily delineated. This provides the opportunity at the drainage basin scale to observe and monitor system operation and change in relation to development, planning and resource use.

- As *open systems*, drainage basins function as a transport mechanism for water and sediments. Contemporary ideas regarding drainage basin management focus on *sustainable management* preserving this natural system function as much as possible. This is in sharp contrast to the 'over engineering' approach of the past where control and regulation of the system was seen as desirable.

- The path, and speed, of water transport through the drainage basin illustrates the interaction between the many surface and sub-surface elements within the basin. The time lag before precipitation reaches a river channel clearly shows the varied constraints on water flow of these interactions. *Direct precipitation* has no constraint on flow to the river part of the system. Vegetation slows transfer to the surface by *interception* and *throughfall*. Ground surface properties promote overland *flow* (faster) and *infiltration* (slower). Transfer downslope through the soil may be either by *interflow* (slower) and *throughflow* (faster). Whilst percolation into suitable underlying rocks (*aquifers*) will recharge the groundwater store.

- The nature of the *discharge* of the river channel system will reflect the various contributions of these flows through time and space. These usually are annual cycles linked to the seasonal climate system or short-term fluctuations reflecting extreme weather conditions superimposed on the annual pattern.

- River networks often reflect the underlying geological control on rates and patterns of erosion. The study of river networks has applications related to the river flood hazard. Initial classification, using a stream ordering system, allows comparison between drainage basin networks. Thus the network system response to precipitation may be monitored in relation to drainage basin properties, such as surface material, vegetation cover and slope, in one catchment and the system data used as a basis for drainage basin management elsewhere.

- The river system has great influence on the surface of the Earth. It is the most important agent of erosion in temperate latitudes, transporting weathered material from the land surface via its network to the sea, promoting landscape change. The processes of transportation vary in effectiveness through time and are dependent on both particle size and velocity. If both variables are not matched for entrainment then the material remains in the network store.

- As a system, rivers often display characteristics of *dynamic equilibrium,* attempting to balance erosion and deposition. This is reflected morphologically in the channel geometry, which shows adjustment to increasing discharge, itself a product of greater catchment area (applicable for most rivers). Consequently, there is a pattern of channel geometry change from the source areas to the mouth of a river network.

- River channel planforms, meanders and braided reaches, reflect the influence of the slope component of the system. There is a critical threshold of slope below which the river channel system responds by meandering. This is a dynamic response to water and sediment flow through the system with many types of evidence illustrating spatial response over fairly short time-scales.

- The slope of a river channel is the most stable aspect of channel geometry. Whilst the ultimate control is the downcutting to sea level, in most instances the geological control of resistant rock or human interference such as dams exercises a practical control over this type of system adjustment.

- All the geometrical components (planform, cross-profile and slope) are dynamic attempts to keep the system in equilibrium. They are observed as changes at different time-scales. Over the short term channel changes are controlled by the discharge regime as a product of weather. Extreme events producing large flows will cause dramatic channel change. At the medium term (thousands of years) the regional climate system is the dominant control. Over timescales longer than this uplift from the tectonic system provides overall control.

- Deltas are at the interface of the surface hydrological system and the marine component of the hydrosphere. They reflect depositional response to the particular processes operating at this interface and the nature of the surface morphological graduation from land to sea.

Main learning hurdles

Downstream velocity change

Students sometimes fail to appreciate the subtle balance of factors controlling water velocity in a channel. They tend to concentrate on the long profile, seeing slope as the dominant factor. The instructor must emphasize the importance of friction. Bed and bank friction can be tied in with size of channel in relation to the spatial contribution from the drainage basin at various points and the need for the river to transport increasingly larger volumes of water downstream. Internal friction may be ably demonstrated in the field by observation of eddies and flow patterns in steep headwater streams. The erratic routes taken by floats should visually convince them of factors at work other than slope. Ensure that observed flow conditions are related to bankful potential, illustrating the changing relation of friction with discharge volume.

Stream ordering systems

This concept commonly causes confusion if students have been exposed to other methods of stream ordering apart from Horton's. It might be a useful discussion topic to debate the logic behind the schemes of, e.g., Horton and Strahler and evaluate their comparative merits.

Key terms

Aquifer; bedform; braided; channel bars; channel network; closed system; competence; conduits; deltas; dendritic; deposition; discharge; drainage basin; evaporation; fluxes; groundwater; hydrology; hydraulic efficiency; integrated river management; interception; laminar flow; lateral accretion; long profile; meanders; misfit streams; nested hierarchy; overbank flow; ox-bow lake; pools and riffles; precipitation; river regulation; runoff; sediment entrainment; sinuosity; stores; sustainable management; Tennessee Valley Authority; terraces; throughfall; turbulent flow; velocity; wavelength.

Issues for group discussion

Discuss the human influence on river systems

The students should read the basic reviews of Park (1981) and Gardiner and Park (1978) to provide the baseline theory and topical content. The instructor should limit the discussion to channel management and interference, as the wider implications of drainage basin management will build on these in the chapter on water resources. A review of the impact of the Aswan High Dam Scheme on the River Nile and its delta would be a good bridging session.

Discuss the implication of climate change on river systems

The students should find information on the basic system functions from Chahine (1992); these are formalized as a model in Rind et al. (1992). From this material the students should be able to suggest their own scenarios. This may then be developed by comparison with the work of Newson and Lewin (1991). A key area for the instructor to look out for in the discussion is the interdependence of climate and surface river systems.

Selected reading

Chahine, M. T. (1992) 'The hydrological cycle and its influence on climate'. *Nature* 359 (6394), 373–80.
A clear review of the hydrological cycle in relation to the interaction of the hydrosphere, atmosphere and biosphere. Emphasizes the need for multidisciplinary teamwork to improve understanding of the complex cycling of water. Useful for explaining terminology in system format.

Gardiner, V. and Park, C. C. (1978) 'Drainage basin morphometry: Review and assessment'. *Progress in Physical Geography.* 2 (1), 1–35.
A key review explaining the main concepts of drainage basin morphometry. It clearly links these in a systems framework to a variety of applications. Suggestions for future management uses are laid out.

Newson, M. and Lewin, J. (1991) 'Climatic change, river flow extremes and fluvial erosion-scenarios for England and Wales'. *Progress in Physical Geography* 15 (1), 1–17.
Focuses on the management responses to future climatic change. Critically examines the ability of hydrological system modelling to predict future scenarios. Concentrates on system dynamics with a strong emphasis on response mechanisms to a variety of changes affecting the hydrological cycle. Future policy is seen to be necessarily holistic in outlook based on a comprehensive knowledge of system workings at a variety of scales.

Park, C. C. (1981) 'Man, river systems and environmental impacts'. *Progress in Physical Geography* 5 (1), 1–31.
A summary of the main human impacts on fluvial processes and resultant landforms. Explores the multidisciplinary nature of the area and the wide spatial extent of the issues. Systems knowledge and application is again seen as a key feature of sustainable management.

Rind, R. C., Rosenzweig, C. and Golberg, R. (1992) 'Modelling the hydrological cycle in assessments of climate change'. *Nature* 358 (6382), 119–22.
Models that predict hydrological responses to global warming are critically examined. Imperfect system knowledge compromises their use in predicting impact on water circulation and availability, with ramifications for agriculture and forestry systems.

Textbooks

Knighton, D. (1984) *Fluvial Forms and Processes.* Edward Arnold: London.

Good intermediate text that graphically displays data to illustrate fluvial relationships.

Newson, M. (1994) *Hydrology and the River Environment*. Oxford University Press: Oxford.
An introductory text that adopts a multidisciplinary stance. The basic system of the water cycle is clearly explained with many illustrations. Basic scientific evidence and methods are used to explore applications in relation to environmental issues concerning the hydrosphere.

Shaw, E. M. (1994) *Hydrology in Practice*, 3rd edn. Chapman and Hall: London.
An advanced text that details hydrological processes in scientific language. Assesses a variety of applications to which these hydrological principles may be applied.

Essay questions

1 Examine the characteristics and the causes of either braided or meandering channels.
2 Outline the variables, and their functions, that affect river channel form.
3 Discuss the role of bank material in determining river channel pattern.
4 With reference to river systems, discuss the importance of time in geomorphological explanation.
5 Discuss the circumstances that favour overland flow.
6 Evaluate the impact of forestry on the hydrology of a drainage basin.
7 Describe the relationships between velocity and sediment size, and erosion, transportation and deposition.
8 The drainage basin may be considered as a number of stores interlinked by various transfer processes. Describe how these stores and transfer processes influence either water or sediment output from the drainage basin system.
9 Examine the effects of three types of engineering works on channel form and process.
10 Critically analyse the controls on delta formation and growth.

Multiple-choice questions

Choose the best answer for each of the following questions.

1 Most of the water held in storage within the hydrological cycle is in:
(a) clouds
(b) the oceans   *
(c) the Antarctic
(d) reservoirs

2 How much of the total water in the global hydrological cycle is freshwater?
(a) 60 per cent
(b) 20 per cent
(c) 2 per cent   *
(d) 1 per cent

3 Which of the following natural water stores has the greatest residence time range:
(a) rivers
(b) rock   *
(c) ice
(d) plants

4 A drainage divide is also called a:
(a) watershed   *
(b) tributary
(c) slope
(d) basin

5 TVA stands for:
(a) Tennessee Valley Association
(b) Tennessee Valley Authority   *
(c) Total Valley Administration
(d) Tennessee Valley Administration

6 Direct precipitation is:
(a) water piped from a reservoir
(b) water flowing downhill
(c) rainfall in urban areas
(d) water falling without stores   *

7 Which of the following contributes to overland flow?
(a) a sloping surface
(b) a saturated soil
(c) an impermeable soil horizon
(d) (a), (b) and (c)   *

8 Rocks that allow water to pass through them are:
(a) sedimentary
(b) phreatic
(c) permeable   *
(d) porous

9 Which of the following is a stream ordering system?

(a) Houghton
(b) Dendritic
(c) Horton *
(d) Lawton

10 How many of the world's ten longest rivers are in Asia?
(a) 5 *
(b) 3
(c) 2
(d) 1

11 The Hjulstrom graph illustrates that:
(a) smallest and largest particles require higher velocities for entrainment
(b) higher velocities are required for entrainment, for a given size of particle, than for keeping it in motion once entrained
(c) when velocity falls below a critical level for a particle deposition occurs
(d) (a), (b) and (c) *

12 What is likely to cause a waterfall?
(a) a widening of the channel
(b) a hard rock outcrop *
(c) flood flow
(d) an increase in velocity

13 Change in channel slope may be determined from the:
(a) cross profile
(b) Hjulstrom graph
(c) stream hydrograph
(d) long profile *

14 Which of the following is not a component of stream load?
(a) reload *
(b) suspended load
(c) dissolved load
(d) bedload

15 Rocks that store water are called:
(a) caverns
(b) sinkholes
(c) aquifers *
(d) aquicludes

Figure questions

1 Figure 11.4 shows the water table and associated slope section. Answer the following questions.

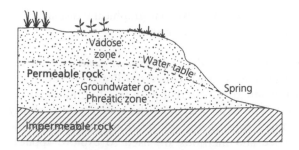

Figure 11.4   The water table. The water table is the upper level of groundwater in permeable rocks beneath the ground surface, below which the rocks are saturated. After Figure 1.9 in Buckle, C. (1978) Landforms in Africa. *Longman, Harlow*

(a) Define groundwater flow.
(b) In relation to other types of flow discuss its contribution to storm flow from a drainage basin.

Answers

(a) Groundwater flow takes place within the bedrock. Above, the water table water percolates through the vadose (unsaturated zone) to the groundwater (saturated zone). Groundwater flow takes place within the groundwater zone to the spring.

(b) The relative importance of flow contributions to stormflow from a drainage basin may be defined in decreasing importance from the surface down to the groundwater source. Overland flow provides the fastest response to the channel from storm conditions where precipitation rates exceed infiltration capacity of the surface material. Throughflow within the soil layer contributes quite rapidly as well, particularly if the rainfall is recharging saturated soils adjacent to river channels. Groundwater flow is generally too slow to contribute to these flow conditions. However, with thin soils and a high water table it is difficult to determine when exactly throughflow ceases and groundwater flow contributes.

2 Figure 11.5 shows a storm hydrograph. Answer the following questions.
(a) Describe the form of the graph.
(b) What factors may affect the lag time?
(c) How might the rainstorm affect channel form?

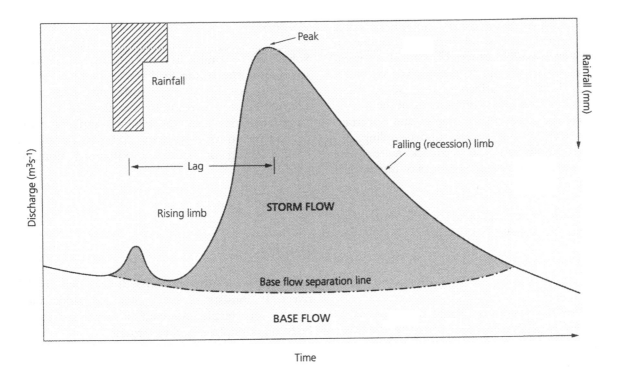

Figure 11.5 *The storm hydrograph. A storm hydrograph displays the variations in river discharge through time, in response to the rainfall input by an individual storm event. Storm flow is the rapid runoff related directly to the storm input.*

Answers

(a) There is a small peaked response related to direct precipitation input, followed by a rising limb which reflects quickflow (overland flow). The time lag between the flood peak and intense rainfall reflects time of travel to the channel. The recession limb falls away, as the discharge drops, with throughflow contributions. Eventually the curve flattens out as the river returns to base flow conditions.

(b) Lag time will be influenced by the channel network density and the shape of and size of the drainage basin. High network density reduces the average surface distance overland flow needs to travel to contribute to the river channel. Small drainage basins respond more rapidly for similar reasons whilst basins with a more uniform circular shape contribute to the main channel from various parts of the drainage basin at a similar time. The more elliptical a basin becomes the longer the lag time.

(c) Both increased sediment and water input to the channel may alter its form. Severe overland flow may carry debris to the channel which under flood conditions will scour the bed and erode the banks, effectively increasing channel cross-section. Increased discharge will erode the banks up to bankful capacity

effectively widening the channel. Movement of bedload under these conditions will erode the bed and also redeposit material downstream.

Short-answer questions

1 List the main sets of processes that drive the hydrological cycle.

Answer

The hydrological cycle is driven by: evaporation and transpiration (movement of water from the surface to the atmosphere); precipitation (movement of water from the atmosphere to the surface); and air movement which redistributes water within the atmosphere.

2 Outline the uncertainty surrounding possible links between the hydrological cycle and global warming.

Answer

Climate warming will affect the hydrological cycle via changes in the relationships between the fluxes and stores, such as cloud cover, atmospheric water vapour concentrations, wind patterns, evaporation rates and condensation processes. Global warming will, at least partially, melt the global ice store, raising sea level and affecting the global water balance. As it affects most components of the hydrological cycle the situation is complex and uncertain given our lack of full system understanding. Of greater uncertainty are the feedback mechanisms that will operate between these two intrinsically linked systems once one changes.

3 What values are expressed by the sustainable management of river systems?

Answer

There are several key areas related to a sustainable approach to river management. Initially the river must be seen as an integral factor in general system operation, such as its need within ecosystems and its role in global biogeochemical cycling. The river must be viewed as a multiple resource for a range of human activities. Our custodianship of the river water must focus on reducing both demand and waste. Reducing the input of pollution and centralizing the importance of river systems within overall sustainable development strategy are also immediate goals.

4 List the factors that control river flow.

Answer

Many different factors affect river flow but the main controls are: area of the drainage basin; precipitation totals; precipitation intensity; and the surface conditions within the drainage basin.

5 What is ephemeral flow?

Answer

Ephemeral flow is when a river flows only at certain times of the year or during and soon after intense rainfall. It is common in semi-arid areas.

6 Describe the three main ways rivers transport solids.

Answer

Solids may be soluble in water and are transported in solution as the solute load. Small particles may be held in suspension by turbulent eddies and transported downstream by the flow of the river. Larger particles form the bedload and may move, dependent on river velocity and particle size, by rolling/sliding or bouncing (saltation) along the river bed.

7 What is alluvium?

Answer

Alluvium is the fine-grained sediment deposited by rivers. It consists of mud, silts and sands and is usually very fertile. Alluvium is deposited in river beds, on floodplains and in estuaries.

8 Outline the relationships found between channel variables and meanders.

Answer

Strong relationships have been found between meander wavelength and channel width. This is a positive relationship, with wider channels having larger meanders. Riffle spacing also increases with channel width. With regard to flow, higher discharge promotes longer meander wavelengths.

9 What is equilibrium in relation to rivers?

Answer

A river is in equilibrium when the energy provided to transport sediment and shape the landscape is equal to the work that has to be done. Rivers continually adjust, promoting a dynamic equilibrium within the system threshold.

10 Suggest factors that might influence river terrace development.

Answer

Base level may be lowered due to a local fall in sea level. There may be a general uplift of the land affecting the drainage basin, increasing river slope and thus velocity. Human influences might include reduced sediment load due to river impoundment behind a dam and increased river flow due to land-use changes such as deforestation. Increased river flow may also come from natural sources, e.g. glacial meltwater.

Additional references

Brammer, H. (1996) 'Bangladesh's braided Brahmaputra'. *Geography Review* 10 (2), 2–7
System stability is explored in relation to seasonal change. Spatial results of a variety of causal factors are clearly illustrated, with emphasis on management intervention at the drainage basin level.

Collier, M. P., Webb, R. H. and Andrews, E. D. (1997) 'Experimental flooding in Grand Canyon'. *Scientific American* 276 (1), 66–72.
Describes the channel changes associated with a controlled flood event. Spatial movement of sediment is clearly described and illustrated. Management applications relating to river channel restoration and enhanced tourism value are evaluated.

Gregory, K. and Branson, J. (1996) 'Back to the future: Rivers past, present and future'. *Geography Review* 10 (2), 28–33.
Contemporary measurement and past records are discussed as predictive tools for future drainage response to climate change. The dynamism of drainage basin systems is emphasized with responses in both channel flow and form. Palaeohydrology as an applied area of study is positively considered.

Higgitt, D. (1996) 'Where the river runs deep'. *Geography Review* 10 (1), 13–15.
A brief review of the processes involved in producing river channel form.

Newson, M. (1996) 'Catchment plans: A new geographical resource'. *Geography Review* 9 (3), 17–24.
An article that focuses on the drainage basin as a logical unit for holistic environmental management. Outlines the criteria assessed and the process of consultation required to aid decision making. Well illustrated with flow charts and a variety of maps to emphasize the spatial interaction of a catchment's sub-systems.

Web site

www.igc.apc.org/green/green
A site that concentrates on drainage basin systems. Material presented may be rather variable at this level.

Aims

- To identify the water resource and examine the nature of demand for water.

- To look critically at human modification of drainage basins and their river networks.

- To explore the critical environmental functions associated with some water habitats.

- To outline the problems arising from water resource use and the natural hazards associated with terrestrial water systems.

Key-point summary

- Global water supply has the potential to meet human demand. The problem is the *uneven distribution* of the resource, *unequal consumption* of the resource and the often wasteful inefficient use of water.

- The main demands on the water resource globally are from *agriculture* (70 per cent) and *industry* (23 per cent). These human economic systems impact back on water supplies via *water pollution*. A feedback loop exists whereby water resources attract human activity and this activity adversely affects the quality of the supply.

- To address the problems associated with meeting demand for water, both quantity and quality, co-ordinated long-term planning is required. The difficulties of organizing a coherent *water management plan* are compounded by scale. The larger the scale the more complex the natural system with greater variations between components, such as different inputs from different climatic zones.

- Surface water networks often cross political boundaries. As change to one part of the water system affects other components, usually downstream, then water resource use and management in one country may affect the potential water resource of another country. This potential for conflict can only be alleviated by international harmonized management of the whole catchment network. At a smaller scale conflict arises between sectors of water users, e.g. farmers and city planners. This adds to the complexity of system management beyond the two-dimensional spatial scale to include a third dimension of *multiple user groups* superimposed on this space.

- The underground stores of water should not be ignored when examining the water resource system. Surface activities may affect their quality and over-*abstraction* will threaten their yields both in terms of quantity and quality (mineral salt concentration). The balance between these sub-surface parts of the system and other components is extremely delicate.

- Surface water may be used to provide *hydropower*, the largest renewable source of energy around the world. The interaction of the relief system (topography and slopes) with the surface water system is fundamental to the selection of usable sites. To harness the potential economically, significant modifications to the river channel network are required. These impact back on the system (reservoirs restricting sediment transport) and on other systems (displacement of human settlement). These widespread effects on other natural and human systems will probably constrain the increasing implementation of hydropower schemes in the future.

- At a global scale the single greatest use of water is for cropland irrigation. This demand is growing rapidly. Again, modification of the river network system causes both benefits (*increased crop yields*) and problems (*salinization* and *waterlogging*).

- To exert control over the supply of water both through time and space often requires regulation of the natural system flow. *Large dam schemes* enable this. The benefits of consistency and timing of flow accrue but there are many problems related to a change in water distribution status. Flows, stores and processes within the water cycle are affected with subsequent effects on the other major environmental systems, e.g. *seismic stress, habitat loss* and *nearshore flow circulation*.

- Too much water may interfere with the functioning of other systems. *Waterlogging* affects plant growth, via the soil system, and agricultural yields when added inappropriately, often due to poor knowledge of system processes (drainage) and their function. However, habitat areas evolve where there is a natural surfeit of water, e.g. *wetlands*. In the past they have often been viewed as a wasted resource, being either *infilled* or *drained*. More recently, with a greater understanding of their environmental value, they are better protected.

- Saline water, whilst being a key component of the marine system, is misplaced and damaging in the terrestrial environment. It is a water property that is not fixed, so salination (increased salinity) on land or in freshwater ecosystems is a negative impact. It is a natural process either by marine intrusion into coastal aquifers or by evaporation on land. However, human actions may intensify the process through irrigation schemes. Desalination is a completely artificial process providing fresh water. It is not without environmental impact, being energy intensive and producing concentrated brine as waste.

- Water pollution, like air pollution, is a widespread problem due to the ease of transfer along system pathways. Severe water pollution has the potential to affect all species, including humans. Water pollutants may reach critical concentrations and become toxic to life. This may occur due to a short-term concentrated release into the water system or long-term accumulation in a store, e.g. groundwater. Water pollution sources may be areal in extent, such as fertilizer use leading to *eutrophication*, or a point source such as industrial effluent piped into a stream. Regional pollution may occur from point sources by means of indirect routes via other systems, e.g. industrial point sources inputting into the atmosphere and acid rain falling over a wide area, polluting a variety of water stores, again via runoff from the land system. With such complex linkages it is difficult to manage water pollution once it gets into the water system. The key management measure, therefore, has to be the pre-treatment or non-release of pollutants into the water system.

- Eutrophication is an example of the human agricultural system interacting inappropriately within the hydrological cycle to produce water pollution problems. Following widespread human input (often encouraged by politico-economic policy) fertilizers are transported from the land to freshwater systems by natural processes. This nutrient enrichment causes a system response, encouraging the growth of bacteria and algae, which produces an imbalance in the water environment.

- *Floodplains* are a natural spatial component of the drainage basin system. They contain a variety of land resource ranging from fertile agricultural land to swampy wetlands. This often causes conflict between economic and environmental interests in relation to both pollution and land drainage.

- Whilst a *flood* reflects an inability of drainage basin controls to manage the transport of an intense input of water into the system over short time periods, the overall equilibrium is usually quickly restored afterwards. This is a good example of the natural fluctuations around a long-term trend line (dynamic equilibrium). Flood analysis of magnitude and frequency is used for managing the hazard. The usual response has been to change system components and flows by physically altering (engineering) the river system, though a more passive sustainable approach in sympathy with the natural processes of drainage basins is now promoted.

Main learning hurdles

Environmental pathways

A revision of the hydrological cycle from Chapter 11 is appropriate in order that the students appreciate the system links facilitating pollutant movement. They should be especially clear regarding the land to river system flows, and the groundwater store.

Eutrophication

Students with a limited scientific background need to be aware of the relationships between nutrients and productivity in aquatic systems. The importance of maintaining aquatic life needs to be clarified.

Key terms

Acidification; artesian well; dams; drainage; Environmentally Sensitive Areas; eutrophication; fertilizers; floods; groundwater abstraction; hydropower; hydrostatic pressure; inefficiency; inequality; irrigation; leaching; levees; Mar del Plata Action Plan (MPAP); nitrates; pesticides; phosphates; pollutants; salination; Somerset Levels; toxins; waterlogging; water table; wetlands.

Issues for group discussion

Discuss the costs and benefits of the use of small-scale hydropower schemes

The instructor should encourage the students to consider the wide range of factors involved: water supply; climate; sedimentation; and the political and population structure. Background and rationale for micro-hydropower schemes are provided in the chapter and larger-scale case studies may be evaluated from Park (1980) and Macmillan (1989).

Discuss the contention that water should be traded as a commodity like other economic resources

The discussion might initially focus on the distribution of freshwater resources, with particular emphasis on arid and semi-arid areas. It should be broadened out to look at the 'ownership' of freshwater resources and the potential political conflict that may ensue. Finally, the concept of global sustainable development should be explored regarding the feasibility of providing water for all.

Discuss the management of the flood hazard

Students should read Smith (1993) to gain an overview of the topic. Development of the discussion can usefully relate to systems and scale. Examples at the small scale, such as Magilligan (1985), and the large scale (Knox (1993)) should provide evidence of the complexities involved in floodplain development. They also highlight anthropogenic activities and the control of these will provide a sympathetic management focus in comparison to engineering schemes.

Discuss the relative threats to wetland areas and their management

The discussion should be started by examining the fragility and importance of such systems, e.g. Brooke (1990). Particular threats may be explored via Trudgill (1989) and Heathwaite (1994). In summation the ideas of the group towards management may be compared to the case study area reviewed by Terry and Case (1994).

Selected reading

Brooke, J. (1990) 'A positive future for Broadland?' *Geography Review* 3 (3), 31–6.
The holistic use of a major wetland habitat is examined. Emphasizes the fragility of these ecosystems and how management may be applied to preserve the systems' integrity. Usefully illustrated with some good flow diagrams.

Heathwaite, L. (1994) 'Eutrophicaton'. *Geography Review* 7 (4), 31–7.
Demonstrates the link between land and aquatic systems. The effects of human systems in relation to natural chemical cycles are discussed in a time sequence framework. Evidence from this is used to discuss management strategies.

Knox, J. C. (1993) 'Large increases in flood magnitude in response to modest changes in climate'. *Nature* 361 (6411), 430–2.
Uses past evidence to suggest that small climatic changes due to greenhouse gas-induced warming will produce magnified outcomes in other systems. Large regional-scale river network systems would breach water transport thresholds and large-scale flooding would result.

Macmillan, B. (1989) 'The Three Gorges Dam: Practical geography on a grand scale'. *Geography Review* 3 (2), 36–40.
Applied spatial, environmental decision making is used to evaluate this case study. A full range of human and physical systems is analysed regarding the pros and cons of the project. Clearly illustrates the multidisciplinary nature of the study of environmental impacts.

Magilligan, F. J. (1985) 'Historical floodplain sedimentation in the Galena River Basin, Wisconsin and Illinois'. *Annals of the Association of American Geographers* 75 (4), 583–94.

This article relates human impact on land cover to the process of floodplain sedimentation. Flood events and their rapid response in humid regions are discussed in a well-illustrated paper.

Park, C. C. (1980) 'The Grande-Dixence hydro-electric scheme, Switzerland'. *Geography* 65, 317–20.
A case study of the rationale behind dam construction, its aims and an outline of impacts.

Smith, K. (1993) 'Riverine flood hazard'. *Geography* 78 (2), 182–5.
A short review assessing the variability of flood risk. Strategies for reducing flood losses are discussed with particular reference to urbanization.

Terry, A. and Case, D. (1994) 'Management of the Somerset Levels and moors'. *Geography Review* 8 (2), 13–17.
Clearly illustrates the relationship between wetland habitat management and the ability to support floral and faunal diversity. The case study is examined in relation to environmental policy criteria.

Trudgill, S. (1989) 'The nitrate issue'. *Geography Review* 2 (5), 28–31.
An outline of the main issues relating to the use of nitrate fertilizer. Clearly demonstrates pathways between physical systems and human systems. NB: The diagram of the nitrate cycle given here is simplified; a comprehensive diagram is included in the January 1990 issue (p. 27).

Textbooks

Barnes, R. S. K. and Mann, K. H. (1991) *Fundamentals of Aquatic Ecology*. Blackwell: Oxford.
Provides a comprehensive coverage of the functioning and value of a range of water habitats.

Gleik, P. H. (ed.) (1993) *Water in Crisis: A Guide to the World's Fresh Water Resources*. Oxford University Press: Oxford.
This guide focuses on the severity of the global freshwater crisis. It is extensively illustrated providing the instructor with a wide range of data sets.

Gupta, A. (1988) *Ecology and Development in the Third World*. Routledge: London.
A sound appraisal of the threats to water resources. Includes useful chapters on air pollution and forestry with implied links to the hydrological cycle.

Johnson, D. L. and Lewis, L. A. (1995) *Land Degradation: Creation and Destruction*. Blackwell: Oxford.
Relates the impacts of land abuse to a range of water resource issues. Includes good system links between water quantity and quality status.

Lidstone, J. (ed.) (1995) *Global Issues of our Time*. Cambridge University Press: Cambridge.
A range of discrete topics is presented with relevance to this chapter. There is explicit focus on the impacts of dams and also the flood hazard. Links to the water resource from soil erosion and acid deposition are also included.

Mitchell, B. (1990) *Integrated Water Management*. Belhaven: London.
A comprehensive review of the issues of managing the water resource with a strong emphasis on 'system thinking' for sustainable management.

Thanh, N. C. and Biswas, A. K. (1991) *Environmentally Sound Water Management*. Oxford University Press: Oxford.
An overview of the practices and principles of sustainable water resource management. These are discussed in relation to 'real world' policy-making decisions.

Essay questions

1 Evaluate the human processes that influence the flood hydrograph in large drainage basins.

2 Outline and justify the treatment of drainage basins as transfer systems.

3 Summarize the main pollution burdens and their effects on water quality within a drainage basin.

4 With reference to examples, contrast the uses of water by traditional and intensive farming methods.

5 Describe the major stores of water within a drainage basin and evaluate their potential for sustainable management.

6 In what ways may runoff from an urban area affect the chemical and biological quality of a river?

7 At the local scale describe the range of management practices that may enhance the conservation value of a wetland area.

8 Evaluate the environmental problems associated with irrigation schemes.

9 What are the possibilities for increasing water supplies in semi-arid areas?

10 To what extent, and in what ways, might environmental management alter the nature of the flood hazard?

Multiple-choice questions

Choose the best answer for each of the following questions.

1 Agricultural use accounts for how much of the global water demand?
   (a)   7 per cent
   (b) 23 per cent
   (c) 50 per cent
   (d) 70 per cent   *

2 By how much more does a typical person in the West use water than a typical person in India?
   (a) 2,000 per cent
   (b) 1,200 per cent   *
   (c)   800 per cent
   (d)   200 per cent

3 How much of the world's electricity is supplied by HEP?
   (a) 20 per cent   *
   (b) 15 per cent
   (c) 10 per cent
   (d)   5 per cent

4 Irrigated land comprises approximately how much of the world's farmland?
   (a) 15 per cent   *
   (b) 25 per cent
   (c) 35 per cent
   (d) 45 per cent

5 Salinity is:
   (a) a result of severe problems
   (b) a cause of eutrophication
   (c) a serious problem in irrigated cropland   *
   (d) a measure of acidity

6 Waterlogged soils are usually a product of:
   (a) good drainage
   (b) soil erosion
   (c) cumulative processes   *
   (d) infrequent large-scale processes

7 The major salt component of seawater is:
   (a) magnesium sulphate
   (b) calcium sulphate
   (c) sodium chloride   *
   (d) magnesium chloride

8 The Camelford water pollution incident concerned an input of:
   (a) radioactive waste
   (b) aluminium sulphate   *
   (c) lead
   (d) copper sulphate

9 The washing of nitrate and phosphate fertilizers into a stream causes:
   (a) sedimentation
   (b) oxidation
   (c) a growth in fish size
   (d) eutrophication   *

10 A tributary flowing alongside the main river channel due to obstruction by levees is called:
   (a) a Kazoo
   (b) a Back swamp
   (c) a Cajun
   (d) a Yazoo   *

Figure questions

1   Figure 12.9 illustrates the ecological impact of oxygen depletion in a river. Answer the following questions.
   (a) Define the biochemical oxygen demand.
   (b) Describe the trends downstream of the two graphed variables.
   (c) What is the river life response to the downstream changes.

Answers

(a) Dissolved oxygen is the amount of oxygen absorbed by the water from the air and available to aquatic life. Water at lower temperatures will contain more dissolved oxygen. Biochemical oxygen demand is the oxygen consumed by organic pollution in the water.

(b) Dissolved oxygen is initially at a high level, beginning a downward slope just upstream of the effluent input. This slope bottoms out at the start of the septic zone and recovers at a similar slope to attain its previous level by the clean zone. A vertical rise in BOD adjacent to the effluent is experienced which accounts for the oxygen sag. The high level of BOD is attained throughout the decomposition zone, where towards the end it begins a downward slope and curve mirroring the oxygen curve's upward slope. The cross-over between the two is at the start of the recovery zone. The relationship between the two reflects the depletion of oxygen by organic pollution and the self-purifying (dilution) nature of the river as it moves downstream.

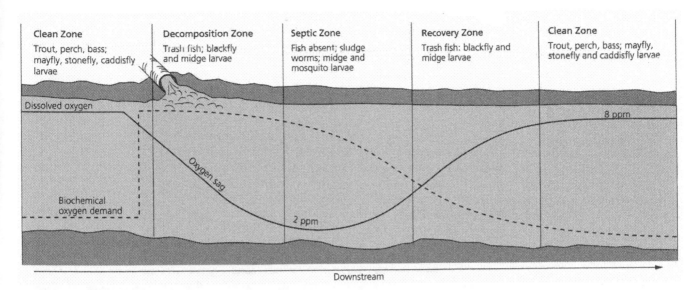

Figure 12.9 Downstream ecological impact of oxygen depletion in a river. Oxygen depletion caused by increased biological activity downstream from a pollution source can radically alter the species composition of the river system. After Figure 22.5 in Cunningham, W.P. and B.W. Saigo (1992) Environmental science: a global concern. *Wm.C. Brown Publishers, Dubuque*

(c) As the oxygen curve sags in the decomposition zone non-scavenging fish and macroinvertebrates, such as mayflies, which are sensitive to oxygen levels disappear. In the septic zone, as oxygen continues to drop, all fish and the flies disappear, with sludge worms thriving on the effluent. Once the dissolved oxygen level becomes greater than the BOD then the species composition recovers to a similar make-up to that in the decomposition zone. Further downstream normal levels are found and the species composition retains its initial status.

2   Figure 12.10 shows the multiple sources of water pollution. Answer the following questions.
   (a) How can changes in the vegetation cover affect the input of pollution to the river system?
   (b) Outline the changes through time of the agricultural input.
   (c) Describe the role of clouds in the water pollution system.

Answers

(a) Vegetation provides a protective cover for the land surface. Removal of the vegetation will leave the bare soil exposed to both wind and water erosion, inputting sediment to the river system.

(b) The agricultural input will change throughout the year due to changes in chemical input, e.g.

fertilizers, at different stages of the crop cycle. Other inputs, dependent on agricultural practices, may be organic (manuring) or particulate (soil cultivation).

(c) Clouds are the stores prior to precipitation. They hold the acidic output of industrial pollution and release it as acid rain. Particulates may also be included in this precipitation inputted to the drainage basin system.

Short-answer questions

1 Describe the UNEP conclusions on water management.

Answer

The UNEP has assessed the ability for sustainable water management throughout the world. They have concluded that national sovereignty is paramount and that decisions regarding the best approach should be made by each individual country.

2 Describe the main problems for water resource management.

Answer

Many different types of problems face the water resource manager. Spatial problems exist due to the

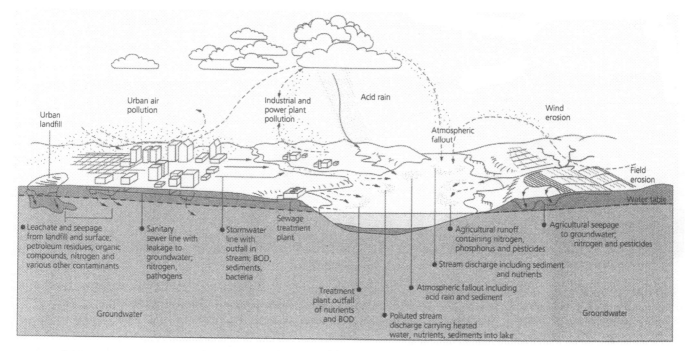

Figure 12.10 Multiple sources of water pollution. Water pollution is generated from many different sources and by many different processes, and land-use changes can significantly alter the amount and type of pollution in a river system. After Figure 11.3 in Marsh, W.M. and J.M. Grossa (1996) Environmental geography: science, land use and Earth systems. *John Wiley & Sons, New York*

uneven distribution of supply and demand. Storage and transfer schemes are needed to redistribute the water. Precipitation is variable through time and again this distribution rarely reflects demand; another storage problem. Finally, the water must be maintained at a quality fit for its intended purpose. This requires treatment and pollution control.

3 Why were water resources seen as important at the Rio Earth Summit?

Answer

Water resources were seen as important as freshwater availability underpins most social and economic activities. Also the global distribution raised concern with increasing deficits in some areas and the emergence of hydropolitics as an area of conflict. Future scenarios suggested one third of the world's countries will face water shortages within a decade. Health issues in relation to water quality and sanitation were seen as key areas to be addressed by the Agenda 21 proposals.

4 What are the site requirements for large hydro-power plants within the river system?

Answer

A hydropower site needs a large area to store water, usually a valley, which can be dammed to facilitate this storage and regulate flow for the HEP production. There needs to be a natural water input either by small headwater streams in upper catchment areas or across a main channel lower down the drainage basin. All schemes require a large amount of land.

5 Outline the uses of wetland ecosystems.

Answer

As natural systems they provide the appropriate habitat for a wide variety of animals, birds and plants. This is their value and use to conservationalists. They are perceived as attractive landscape elements and are used for tourism, especially water-related recreation. Wetlands, when drained, provide good-quality agricultural land and have an economic use to farmers, though this is usually at the expense of the other uses.

6 What are the technologies available for desalinating water?

Answer

Desalination is achieved by a number of processes. Electrodialysis which concentrates salts out of the water. Reverse osmosis which traps dissolved salt as the water is passed through a membrane. Freezing water permits the growth of freshwater crystals which may be extracted. The final method is distillation, when salt-free water vapour is condensed.

7 In the USA what is the major source of water pollution?

Answer

Studies suggest that agriculture is the major contributor. Up to 55 per cent of total river length studied was estimated as being affected by this source.

8 Outline the main efforts made within the European Union to control water pollution.

Answer

There has been an effort to standardize water quality standards throughout the EU. A variety of legislation has been passed to protect aquatic environments, with the use of a variety of economic measures, e.g. fines and incentives to support the legislation. Finally, there has been general encouragement for the introduction and use of more environmentally friendly products.

9 Describe the most important causes of floods.

Answer

Natural causes relate to increased input to the drainage basin system from intense rainfall. Also there may be a rapid release from store into the river system by melting ice and snow. Human stores (reservoirs) may be released suddenly by dam failure. Whilst combined effects such as the displacement of reservoir water, e.g. Lake Vaoint, by a landslide will send a flood surge through the system.

10 What are the applications of flood frequency analysis?

Answer

Applications include land-use zoning in relation to expected flood levels. Also to assess the relevance of engineering works as the most economic way of controlling the flood risk, and if adopting this method the scale of works required to counter the likely magnitude of flood.

Additional references

Brammer, H. (1996) 'Can Bangladesh be protected from floods?' *Geography Review* 9 (4), 21–7.
Clearly illustrates the complexity of river systems in response to flood events. The system scale usefully incorporates a diversity of landform to reinforce the variety of impact both on the natural environment and the people living in it. NB: The relationship of the agricultural system and seasonal flood cycle diagram is amended in the September 1996 issue (p. 33).

Hirsch, P. (1995) 'Sustainable watershed management in southeast Asia'. *Geography Review* 9 (2), 30–4.
Usefully discusses the applicational importance of sustainable development. Outlines the variety of scales of catchment system and uses an international focus. Very useful system diagrams and glossary.

Humberg, C., Ittekkot, V., Cociasu, A. and Bodungen, B. (1997) 'Effect of Danube River dam on Black Sea biogeochemistry and ecosystem structure'. *Nature* 386 (6623), 385–8.
An interesting article on the impacts of large dams on the hydrosphere. Usefully discusses major biogeochemical cycle disruption between adjacent fluvial systems as well as the overall ecosystem impact.

Jones, C. (1997) 'Assessing river water quality'. *Geography Review* 10 (3), 9–11.
A useful article linking physical and chemical components of the river system with species diversity. Clearly illustrates the concept of biotic indicators acting as a synthesis of other environmental conditions.

Middleton, N. (1996) 'The 1995 floods in northwest Europe'. *Geography Review* 9 (5), 25–6.
A brief report that importantly emphasizes the magnitude of natural events. This puts the ineffectiveness of human channel alteration in context at

the large regional scale when flood events reach a certain magnitude.

Patel, T. (1996) 'Bridge over troubled waters'. *New Scientist* 152 (2058), 12–13.
Water rights between India and Bangladesh are explored in relation to engineering schemes. Puts over the issue of 'water wars' and human impact on natural water systems. Policy priorities and compromises from both sides are discussed.

Pearce, F. (1996) 'Water still runs deep'. *New Scientist* 151 (2041), 24–7.
Links the surface water system with climate change. Groundwater storage is seen as the safeguard to supply in response to the increasing drought threat in Britain over the last 20 years. Examines stored distribution in relation to water supply planning and the impact of society on the natural water system.

Pearce, F. (1996) 'Dirty groundwater runs deep'. *New Scientist* 151 (2048), 16–17.
Pollution of groundwater in Great Britain is examined. Illustrates connectivity in the water system, linking surface pollution with sub-surface stores, and those with surface resources. Also discusses economic costs and the future impact of climate change.

Web site

www.nwl.ac.uk/ih/
The Institute of Hydrology site provides case study links to a wide range of water resource projects. A very useful site for illustrating the interactions between major environmental systems.

Aims

- To produce a workable definition of drought and explore its causes.

- To examine the impact of drought on natural and human systems.

- To outline responses to drought by people; how they cope with and alleviate the problem.

- To define spatially, at the global scale, drylands and account for their expansion through time.

- To explore desertification and evaluate it as a serious environmental problem.

Key-point summary

- Drought implies a serious water deficit in an area. This is defined in relation to time (*prolonged period*) and related to abnormal climate conditions for that area (reduced precipitation). Within a long-term climatic cycle droughts appear to occur on a regular basis.

- For management purposes the climatic conditions leading to drought need quantifying, e.g. 75 per cent of average rainfall. These conditions might be viewed as the threshold for the surface systems affected by the water deficit. The scale of the problem will increase in severity as the *moisture deficit* grows; the reduced precipitation conditions continue; the spatial extent widens and an increasing number of people are affected.

- Many droughts are the result of disequilibrium between the climatic system and the human activities in an area. Vulnerable areas relate spatially to regions of already low natural precipitation defined by the atmospheric circulation system and the ocean–land distribution. Human activities often underpinned by *population pressure* increase this natural vulnerability. Once the drought cycle is initiated positive feedback between human response and regional climatic change deepens the problem. Eventually a new equilibrium is achieved as population pressure declines in response to the severe drought conditions.

- Drought is a progressive hazard following a particular time sequence of system disruption. Its effect on human populations is related to the *water management* strategies in place to cope with it.

- Spatially, droughts have effects beyond the area experiencing a reduction in precipitation. As a contributing area to a *drainage basin system* the drought region will influence the network regime elsewhere, e.g. lower flows. These influences will reduce with distance from the drought area.

- There is an unequal global effect of drought on human societies. Economically developed nations, prior to drought, have manipulated the surface water system to increase reserve capacity (stores) artificially to cope with water deficits over the medium term. Less economically developed countries have financial constraints limiting this large-scale control. Droughts are a much more serious problem here. Some large-scale storage projects have been implemented but in these more climatically vulnerable areas they are far less successful. Small-scale localized projects are more relevant. These usually involve a more sympathetic treatment of the water cycle processes, e.g. tree planting to reduce soil moisture evaporation.

- *Deserts* are areas characterized by a climatic regime where evaporation exceeds precipitation and may therefore be spatially defined using this variable relationship at the global scale. As climate has changed through time desert extent has changed accordingly. Interactions between spatial position, surface topography and wind systems influence the *aridity* of deserts.

- Wind and water processes produce desert landscapes. Time-scale is important in viewing the influence of these. Wind tends to be persistent, continually moulding the landscape. Whilst the influence of water is periodic. Both processes have great effect due to the unprotected (*unvegetated*) land surface.

- Whilst *drought* is commonly definable as a deficiency of rainfall, *desertification* is perceived as a highly complex and variable phenomenon that does not readily fit a simple definition. The important aspect is that both climatic factors and human resource use have caused effects to a lesser or greater degree. However, it is defined as a truly global problem affecting land within but also beyond the tropics.

- The causes of desertification occur over different time-scales: *climatic change* (long and medium term) and human actions (short term) such as *overgrazing*, *overcultivation* and *deforestation*. However, these rarely act autonomously.

- In order to cope with desertification the scientific knowledge of its interacting causal processes needs better understanding. However, what is known of the surface system components contributing to desertification is used to promote sustainable management. This involves *educating* peoplé in marginal areas to adopt both better land-use practice and remedial measures such as *sand/soil stabilization*.

Main learning hurdles

Weathering in drylands

It is important to refer back to Chapter 6 so that the students are made aware of the importance of water in moulding desert landscapes. This apparent paradox causes some students problems, perhaps not so much with infrequent surface flow but with the need for water in salt and insolation weathering processes.

Desertification

Students quite often have a one-dimensional view of desertification as producing a sand dune desert landscape. The instructor should point out that designated deserts are not composed entirely of sand (far from it). A careful appraisal of the range of

factors and outcomes plus an examination of the spatial extent of the problem (Figure 13.13) should give a realistic impression of a range of desertification-affected landscapes.

Key terms

Albedo; aridity; arrayos; bajadas; deforestation; desertification; deserts; drought; dry spell; Dust Bowl; ergs; land degradation; playa; sand dunes; wadis.

Issues for group discussion

Discuss the contention that desertification is a natural phenomenon

The discussion should focus on the views of Forse (1989) and Binns (1990). More complex discussion may follow, based on Goudie (1991) initially, and the comprehensive systems operations described by Phillips (1993). Liverman's (1990) case study can be usefully applied to conclude the discussion in relation to the veracity of our evidence-based scientific knowledge.

Discuss how global warming might increase atmospheric hazards in desert areas

The students should examine the global distribution of deserts along with a world map of severe storm hazards. The discussion should be based on major environmental system interaction and a basic knowledge of why deserts are located where they are, and the mechanisms promoting severe storms. Gray (1990) will produce some clues but the students will have to develop the theme of global warming effects on ocean circulation and the spatial relationship with desert areas.

Selected reading

Atkiner, S., Cooke, R. U. and French, R. A. (1992) 'Salt damage to Islamic monuments in Uzbekistan'. *Geographical Journal* 158 (3), 257–72.
A research-based article illustrating the effect of human systems on natural processes with outcomes affecting the human resource. System interactions are illustrated in concise diagrammatic form.

Binns, T. (1990) 'Is desertification a myth? Geography' 75 (2), 106–13.

A critical attempt to define the concept of desertification. The link between food production systems and their overall environmental context is established in relation to case study evidence.

Forse, W. (1989) 'The myth of the marching desert'. *New Scientist* 121 (1650), 31–2.
Considers how limited scientific knowledge is used to frame the environmental problem of desertification. Explores the subject briefly, using temporal and spatial evidence.

Goudie, A. (1991) 'The climatic sensitivity of desert margins'. *Geography* 76 (1), 73–6
Climatic system trends are correlated to spatial zonation at the regional scale. This short article identifies clear links between the hydrologic and land systems.

Gray, W. M. (1990) 'Strong association between West African rainfall and US landfall of intense hurricanes'. *Science* 249 (4974), 1251–6.
Decade-by-decade data is explored to illustrate the link between summer rainfall in the Sahel and the frequency of the hurricane hazard on the US East Coast. Clearly illustrates atmospheric interaction with the ocean at the continental scale and multiple hazard outcomes.

Liverman, D. M. (1990) 'Drought impacts in Mexico: Climate, agriculture, technology and land tenure in Sonora and Puebla'. *Annals of the Association of American Geographers* 80 (1), 49–72.
An holistic case study of the drought hazard. Historical periodicity of the climate hazard and human development underpins the need for applied data analysis. Spatial illustrations illustrate local disparities in hazard outcomes. Various alleviation techniques and their usefulness are discussed.

Middleton, N. (1987) 'Wind erosion in the Sahel'. *Geography Review* 1 (2), 26–30.
The mechanisms of wind erosion are clearly outlined. Both physical and human factors are examined in this regional case study of desertification. Links to larger systems are emphasized.

Phillips, J. D. (1993) 'Biophysical feedback and the risks of desertification'. *Annals of the Association of American Geographers* 83 (4), 630–40.
An assessment of the stability of dryland environments in relation to system feedback. System diagrams illustrate the feedback loops and a series of cause–effect relationships is identified.

Thomas, D. (1986) 'Ancient deserts revealed'. *Geographical Magazine* 58 (1), 11–15.
A highly readable article outlining the evidence, including remote sensing, for climate reconstruction. Desert climatic conditions and their resultant effect on landform are spatially displayed to illustrate their variability over long time periods.

Thomas, D. S. G. (1993) 'Sandstorm in a teacup? Understanding desertification'. *Geographical Journal* 158 (3), 257–72.
Spatially analyses the extent of desertification at global and continental scales. Relevant terms are defined and the process of soil degradation clearly illustrated in a systems form which emphasizes both physical and human processes.

Textbooks

Beaumont, P. (1989) *Drylands*. Routledge: London.
Outlines the various causes of drought. Contains useful examples of a variety of management schemes to prevent and cope with this hazard.

Glantz, M. H. (ed.) (1987) *Drought and Hunger in Africa*. Cambridge University Press: Cambridge.
A continental appraisal with national case studies of the drought hazard and its impact on economically less developed countries.

Mortimore, M. (1989) *Adapting to Drought*. Cambridge University Press: Cambridge.
This text provides a comprehensive coverage of the drought hazard. There are specific sections on desertification and policy formulation. A good range of case studies is included.

Raynaut, C. (1996) *Societies and Nature in the Sahel*. Routledge: London.
A regional synthesis of environmental and social systems provides an in-depth advanced study of the Sahara Desert and adjacent sub-Saharan Africa. The application of knowledge of these system interactions provides the basis for potential future management options relevant at a global scale.

Essay questions

1 Why is the morphology of slopes in arid areas different to those in humid areas?
2 How may the difficulties of water supply to drylands be overcome?

3 Evaluate the view that the role of water is critical to denudation in arid areas.

4 Is the spread of the desert at its margins a natural or human-induced phenomenon?

5 Account for the distribution of deserts at the global scale.

6 Evaluate the main hazards associated with arid and semi-arid areas.

7 Compare and contrast the Dust Bowl in North America during the 1930s with the Sahel drought of the last three decades.

8 Examine the processes involved in the formation of desert dunes and their variety of resultant forms.

9 What are the major controls producing arid climates?

10 Discuss the main factors causing land degradation in temperate climates?

Multiple-choice questions

Choose the best answer for each of the following questions.

1 Hot deserts constitute how much of the Earth's land surface?
  (a) 30%
  (b) 20%   *
  (c) 10%
  (d) 2%

2 UNCOD stands for:
  (a) United Nations Conference on Degradation
  (b) United Nations Conference on Desertification   *
  (c) United Nations Conference on Deserts
  (d) United Nations Conference on Drought

3 The Dust Bowl was a phenomenon of the:
  (a) 1930s   *
  (b) 1940s
  (c) 1950s
  (d) 1920s

4 The Sahel is a region south of the:
  (a) Namibian Desert
  (b) Kalahari Desert
  (c) Arabian Desert
  (d) Sahara   *

5 One of the most common farming practices leading to desertification is:
  (a) contour ploughing
  (b) overgrazing   *
  (c) intensive cropping of wheat
  (d) slash and burn

6 Which of the following is a direct economic impact of drought:
  (a) population migration
  (b) loss of crop production   *
  (c) salinization
  (d) dust blow

7 Of the world's ten largest hot deserts most are located in:
  (a) Africa
  (b) Australia
  (c) South America
  (d) Asia   *

8 Sand deserts are dominated by:
  (a) wind action   *
  (b) erosional landforms
  (c) salt deposition
  (d) high relief

9 Sand dunes that align themselves parallel to the dominant wind direction are known as:
  (a) seif dunes   *
  (b) star dunes
  (c) beach dunes
  (d) barchan dunes

10 Which of the following relief features is the largest:
  (a) pinnacle
  (b) mesa
  (c) butte
  (d) plateau   *

Figure questions

1   Figure 13.7 is a generalized global map of the major desert provinces. Answer the following questions.
  (a) What is an arid climate?
  (b) Explain the causes of aridity in the largest province.
  (c) Describe and explain the main characteristics of the vegetation of arid areas.

Answers

(a) An arid area is one where the precipitation input is exceeded by losses due to exploration and

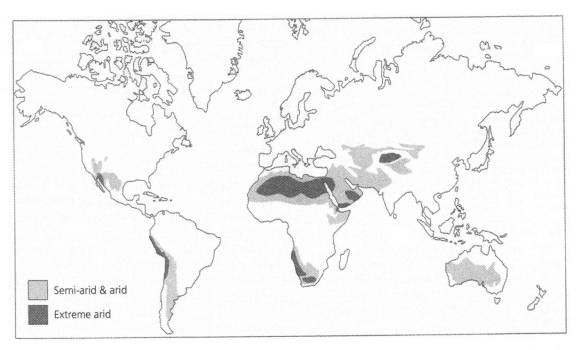

Figure 13.7 *Location of the major desert provinces. Most of the major deserts are located within the tropical climatic zone, but some (such as the Gobi desert in Asia) are located in the dry continental interiors. After Figure 10.2 in Goudie, A. (1993) The nature of the environment.* Blackwell, Oxford

transpiration plus water storage change. Annual precipitation is usually less than 250 mm and is extremely variable. Temperatures are similarly variable, especially the diurnal temperature range.

(b) The largest extremely arid province is the Sahara. The key cause of this aridity is the subtropical high-pressure zone located over the Sahara. The descending air discourages convection and hence precipitation. High insolation inputs in the tropics leads to high evapotranspiration rates. The temperature regime in the Sahara is an extreme one. Summer day temperatures are very high, i.e. 40 degrees centigrade, due to a lack of cloud cover, as a result of low levels of atmospheric water. Thus surface heating from the incoming solar radiation is very effective. Conversely at night the clear skies facilitate rapid heat loss from the ground surface. The variation of temperature is dependent, in the main, on the control exerted on insolation rates by the passage of the Sun between equinoxes.

(c) The lack of moisture and poor soils in arid areas promotes a specialized vegetation cover. Perennial plants are adapted to withstand drought (xerophytes) or have root systems that extend beyond the arid soil zone to tap groundwater supplies (phreatophytes). Xerophytes have low transpiration

rates by closing stomata, having waxy leaves covered with dense hairs and the ability to trap water within their structure. Ephemeral plants are opportunistic rather than adaptive, with seeds lying dormant until sufficient rain allows germination to take place. They grow and flower quickly to provide seeds available for germination in the next wet year.

2 Figure 13.11 illustrates an alluvial fan. Answer the following questions.
    (a) Explain the terms alluvial fan and canyon.
    (b) Why is the stream braided?
    (c) What conditions are conducive to alluvial fan formation?

Answers

(a) An alluvial fan is a fan-shaped mass of rock debris deposited at the junction of a dissected plateau and a plain. A canyon is a deep, narrow steep-sided river valley.

(b) The stream is braided as the alluvial fan has a fairly steep slope, and the size of material is conducive to multi-channel formation.

(c) The conditions needed to form an alluvial fan

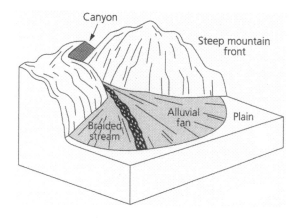

Figure 13.11 Alluvial fans. Alluvial fans accumulate in desert areas where streams flow out from steep-sloped mountain areas, carrying large loads of particulate sediment, much of which is deposited where the stream flows on to the adjacent flatter plain. After Figure 15.14 in Doerr, A. H. (1990) Fundamentals of physical geography. Wm. C. Brown Publishers, Dubuque

are: a fast-flowing stream capable of entraining large rock debris; the opening out of flow from a constricted area to allow deposition due to reduced flow velocity and allowing the stream to divide to produce the fan morphology.

Short-answer questions

1 What were the causes of the Dust Bowl, in the USA, in the 1930s.

Answer

The area affected by the Dust Bowl has a history of persistent drought. Population migration into the area instigated farming on the marginal land. The protective grass land cover was replaced by a crop system that laid the surface open to wind erosion which stripped the dry, easily transported upper topsoil leaving the land severely degraded.

2 What types of drought are there?

Answer

There are four main types of drought. Meteorological drought which defines the rainfall deficit. Hydrological drought produced by reduced river flow. A deficit in soil moisture produces agricultural drought conditions and the overall impact of a deficit in the food resource in relation to the local/regional population is a famine drought.

3 Describe the main resource uses that exacerbate a drought situation.

Answer

Human impact on resources in areas prone to drought further upset the balance of the system and enhance system breakdown. The removal of trees for fuel and building reduces the cohesiveness of the soil and removes surface protection, as does overgrazing which exceeds the carrying capacity of the vegetation system. Inappropriate cropping and water usage systems also aggravate the situation.

4 List the direct impacts of drought.

Answer

Direct impacts include: decreased biological productivity and subsequent lower yields; an increased fire hazard as the vegetation retains less moisture – woody plants are particularly prone; reduced water levels, both in flows and stores; increased mortality amongst animals whose food web is disrupted; and damage to aquatic habitats.

5 Outline the two types of desert.

Answer

The two types of desert are differentiated by temperature as they both experience less than 25 cm of rainfall annually. Cold deserts are found in high latitudes and are covered by snow and ice. Hot deserts are closer to the equator, experiencing higher temperatures.

6 What are the types of surface associated with hot deserts?

Answer

Hot deserts consist of three main surface types. Where sand predominates, forming dunes, the desert is called an erg. If the desert landscape is bare rock it is known as an hammada. A surface comprising loose stones and gravel is referred to as a reg.

7   Describe the effects of dust storms.

Answer

Dust storms are dust-laden winds. They can significantly reduce visibility, dependent on the amount of dust carried, affect communications and threaten agricultural yields by reducing sunlight and coating plant material. When wind velocity is particularly high dust storms have a sandblasting effect on materials. The high particulate concentration in dust storms may cause health problems.

8   List the requirements needed to control desertification.

Answer

Desertification may be managed by: initially being able to detect the onset of desertification by appropriate monitoring; an agreement over causes and severity of process; understanding the appropriate ways of intervening in the process; and a willingness to act quickly before the surface system is so degraded that subsequent management is problematic.

9   Describe the likely impacts facing the Mediterranean region due to projected temperature rises over the next 50 years.

Answer

A rise of 3° centigrade by 2050 will have major impacts on evaporation rates, increasing them by about 200 mm a year. Sea level will rise about 40 cm. Most cereal-growing areas to the north will become abandoned, with a spread of higher-temperature-tolerant species. There will be a rapid increase in desertification triggered by climatic change.

10   What techniques are used in China to counteract desertification?

Answer

The Chinese employ a wide range of techniques based on reducing soil erosion and restoring soil fertility. Wind erosion is reduced by the use of windbreaks; dune stabilization by sediment traps and xerophytic planting plus the general enclosure of land. Soil fertility is restored by irrigating with silt-laden water, the spreading of fertile palaeosols and chemical treatment.

Additional references

Goudie, A., Eckhardt, F. and Viles, H. (1996) 'Geomorphological hazards in the Namib Desert'. *Geography Review* 9 (3), 38–41
Well-illustrated article that introduces a range of hazards associated with extremely arid areas. Discussion of salt, sand and flood hazards only demonstrates the difficulty of inhabiting such areas.

Web Site

drylands.nasm,edu.1995/
This site provides learning activities for student use. Drylands are defined and the threats they face may be interactively examined.

Aims

- To appreciate the impact of past cold climates on the present landscape.

- To examine present cold climates.

- To focus on the processes of glaciation in altering the land surface.

Key-point summary

- The interaction of the climatic system (low temperatures), surface relief system (altitude) and the hydrological cycle produces snow and ice as characteristic inputs and stores within cold environments. Cold environments present many problems and hazards for society due to these conditions. The problems may be exacerbated by other variables, e.g. winds. This climatic severity influences the human system, promoting a distinctive settlement pattern of low-density population with clustered settlements.

- Areas of persistent cold climates are the *tundra* and *polar ice caps*. Tundra is confined to the northern hemisphere due to the land–ocean distribution of the Earth towards the poles. The high latitudinal position exercises control over the atmospheric regime, producing these cold climatic conditions.

- Low general temperatures influence the action of other systems. Precipitation input is mainly in the form of snow. The frozen soil system (*permafrost*) inhibits *percolation* and thus water remains on the surface. This in turn produces problems for society on a variety of time-scales, with permafrost inhibiting engineering and building work throughout the year and seasonal ground instability (summer thawing) producing surface hazards, e.g. mass movement.

- Polar ice caps are the most extreme cold environments with a permanent cover of snow and ice. This reflects the systems balance with a yearly net accumulation of snow and ice. However, due to the low precipitation input this suppresses short-term system overload. Over the shorter time-scale weathering is controlled by temperature change (*freeze-thaw*) and relief.

- In the Arctic the climate system controls the surface water system. Low precipitation reduces river network development and water is mainly held in surface stores (*lakes, ponds* and *marshes*). Human atmospheric influence at the global scale impacts on climatic conditions here via transportation of particulate air pollutants.

- Antarctica, unlike the Arctic ice cap, is a land mass. Its polar position is a result of the plate tectonic system. The remoteness and severe climate make it a unique continent of great importance globally. There has been relatively little human influence (no indigenous population) and it largely remains an undisturbed wilderness. Threats to its integrity are growing and it provides a classic example of the economic versus environmental debate (*exploitation* versus *preservation*).

- Antarctica provides a link to the long-term climatic record via its ice core evidence. Global concentrations of many atmospheric gases and particulates are preserved through time, indicating past events and recent trends.

- *Ice age* conditions are the cold phases in a long-term climatic oscillation. Superimposed over shorter time periods are fluctuations of *glacials* (colder periods) and *interglacials* (warmer periods). The last ice age, the *Pleistocene*, was significant in shaping the landscapes and environments of the mid- to high latitudes that we see today. Within the Pleistocene were a number of glacials and interglacials. These multiple phases of glaciation produced complex landscapes, though the most recent phase is the one that leaves the most surface evidence, having largely removed or remoulded landforms of previous glaciations. The glacial system operates over distinct spatial areas related to low temperatures (high latitudes and altitude) and readvances or retreats according to the general global temperature trend. *Global warming* has the potential to break down the system operation by preventing future advances. This may pose a significant hazard as the melted ice inputs to the hydrosphere, thus raising sea levels.

- The potential causes of glaciation are known but their degree of interaction and the temperatures required to instigate ice advance are not definitively agreed. Geological and scientific evidence suggests this is a sensitive system requiring only a small temperature fall to instigate advance, with positive feedback (*clouds* and surface *albedo*) enhancing the process through time. Over the long term the Earth's variations in orbit control the general temperature trends, whilst smaller system changes such as inputs of volcanic dust, forest fires or air pollution may trigger glacial periods.

- The glacial system is an open system that operates in a state of *mass balance*. In order to advance, inputs (*snowfall*) need to be greater than the outputs (*ablation*). A retreat of the system is the reversal of this situation. The system is spatially controlled with accumulation at the head of the system and ablation towards the lower reaches. An *equilibrium line* differentiates the two areas. At all scales from *continental ice sheets* to valley *glaciers* this balance operates and is essentially controlled by climate. However, smaller glacial systems are more dynamic in their response to temperature change.

- Ice has the ability to shape the Earth's surface by the erosive agents of *abrasion* and *quarrying*. Both processes involve ice movement. These processes remodel existing landforms such as valleys rather than rapidly lowering existing large-scale structures.

- The glacial system is also effective in transporting material and redepositing it to create new landforms. These glacial depositional features are closely related to the system processes forming them and produce evidence of glacial extent and directionality of flow.

- Discrete landforms are also associated with the meltwater output from glacial systems. These *fluvioglacial* features display characteristics of flowing water rather than ice, e.g. stratification. The landforms produced are depositional in process of formation with erosional forms being confined to the subsequent reworking of these by later meltwater flow.

Main learning hurdles

Time-scales of fluctuation

Students sometimes have problems visualizing super-imposed trends and concentrate on glacial and inter-glacial sequences. Reference to Milankovitch cycles (Chapter 7) should be reviewed and the scale sequence of these long-term cycles put in context with glacial/interglacial and stadial/interstadial sequences.

Spatial scales of ice masses

As with time-scales, students should be clear about the different size terminology associated with glaciation. Two features relating to their effects on the Earth's surface often need clarifying. Firstly, large ice sheets, because of their enormous thickness, are often perceived as shaving the land surface of all of its features. This misconception may be tied in with the second difficult area: glacial movement. Not all ice flows at the interface with the surface but may adhere to it, effectively protecting it from sub-aerial erosion. Ice flow is often due to lubrication at the ice–ground boundary, again suggesting a moulding reworking process rather than a complete removal of surface features.

Key terms

Ablation; abrasion; accumulation; albedo; aretes; cirques; climate change; compaction; deposition; drumlins; eskers; firn; fluvioglacial; glacial hazards; glacials; glacial scour; Gondwanaland; Holocene; icebergs; interglacials; kames; kettle holes; mass balance; Milankovitch; moraines; outwash plains; pack ice; permafrost; Pleistocene; polar ice caps; quarrying; Quaternary; seracs; snow and ice; stratified; terraces; till.

Issues for group discussion

Discuss the issue of whether the Antarctic should be preserved or exploited

The students should consider the global importance of 'this last great wilderness'. Discussion should follow the theme of biotic resource. The students should consider not only the impacts within the Antarctic but the wider implications for the planet of the use of the resources. Preservation or conservation definitions may be useful here.

Discuss the contention that the Alpine landscape is primarily a product of glaciation

Students should read Diem (1984) to gain a feel for the complexity of such areas. The instructor should initially focus the discussion on what is landscape and then lead the student to analyse the components of the landscape system and which processes produce them.

Selected reading

Atkinson, K. (1987) 'Life on a deep freeze'. *Geographical Magazine* 59 (9), 444–9.
Human adaptation via technology is investigated in a case study from Arctic Canada.

Diem, A. (1984) 'The Alps'. *Geographical Magazine* 56 (8), 414–20.
An outline description of the contemporary Alpine mountain environment. A variety of patterns produced by both natural and human reactive systems is concisely reviewed.

Perry, A., Symons, L. and Colville Symons, A. (1986) 'Winter road sense'. *Geographical Magazine* 58 (12), 628–31.
Describes monitoring techniques at local and national scales relating to spatial prediction of hazard conditions in winter. Climatic variables and their spatial interaction with human transport networks are illustrated.

Textbooks

Dawson, A. (1991) *Ice Age Earth: Late Quaternary Geology and Climate*. Routledge: London.
Major environmental system change through time is the main focus of this book. Written to appeal at a number of levels, the text overviews interaction producing change between climate, geological and geomorphological systems.

Goudie, A. (1993) *Environmental Change*. Oxford University Press: Oxford.

Directly relates to this chapter in relation to the frequency and magnitude of environmental changes in many systems interacting with climate change in the Quaternary. A wide range of evidence is clearly presented.

Hughes, T. J. (1997) *Ice Sheets*. Oxford University Press: Oxford.
This book examines the dynamics of the atmospheric system that brings about global climatic change. Basic principles are explained and system modelling is used to illustrate the interdependence between ice sheet extent and climate conditions.

Jones, R. L. and Keen, D. H. (1993) *Pleistocene Environments in the British Isles*. Chapman and Hall: London.
A comprehensive account of the glacial environment, its past effects on a range of systems and the contemporary evidence.

Martini, I. P. (1996) *Late Glacial and Postglacial Environmental Changes*. Oxford University Press: Oxford.
A range of evidence is used to reconstruct the rapid changes that occurred during Quaternary glaciation. Usefully suggests applications for the use of evidence in other disciplines.

Essay questions

1 Discuss the erosion processes associated with a periglacial environment.
2 Discuss the evidence for the advance and retreat of a glacier.
3 Describe how large ice sheets influence global climate, and examine the processes involved.
4 With relation to fluvioglacial deposition examine the relationship between sediment characteristics, landform and environment of deposition.
5 Describe the causes and results of variations in the rate of flow of a valley glacier.
6 Should Antarctica be used for tourism?
7 Critically examine the role of meltwater in processes of glacial erosion.
8 Examine the problems of differentiating between the contribution of glacial and fluvioglacial processes in the formation of depositional landforms in the glacial environment.
9 Evaluate the factors controlling polar climates.
10 Examine the concepts of mass balance and the equilibrium line in relation to a valley glacier.

Multiple-choice questions

Choose the best answer for each of the following questions.

1 In the Köppen classification polar climates are:
(a) type F
(b) type E   *
(c) type D
(d) type C

2 Polar climates are defined as areas which have temperatures in the warmest month of:
(a) less than 0° centigrade
(b) less than 5° centigrade
(c) less than 10° centigrade   *
(d) less than 15° centigrade

3 How much of an iceberg mass is below the ocean surface?
(a) 90%   *
(b) 70%
(c) 50%
(d) 30%

4 Antarctica drifted to its present polar position:
(a) 100 million years ago   *
(b) 10 million years ago
(c) 1 million years ago
(d) 100 thousand years ago

5 The average elevation of Antarctica is:
(a) 300 metres above sea level
(b) 1,300 metres above sea level
(c) 2,300 metres above sea level   *
(d) 6,000 metres above sea level

6 The Antarctic Treaty was signed in:
(a) 1959   *
(b) 1969
(c) 1979
(d) 1989

7 In 1991 the Madrid Protocol approved a ban on oil and mineral exploitation in the Antarctic for at least:
(a) 5 years
(b) 50 years   *
(c) 100 years
(d) 500 years

8 Which of the following glacial landforms constitutes a 'basket of eggs' topography?
(a) erratics
(b) kames
(c) drumlins   *
(d) moraines

9 Glacial meltwater streams appear milky due to suspended sediment derived from:
(a) rock flour   *
(b) outwash plains
(c) till
(d) loess

10 Which of the following is a fluvioglacial feature formed below the ice:
(a) kame
(b) esker   *
(c) drumlin
(d) sandur

Figure questions

1   Figure 14.11 shows the basis of a glacier's mass balance. Answer the following questions.
(a) What are the inputs to and outputs from the system?
(b) Why should the mass balance vary spatially?
(c) What would cause temporal variations to the mass balance?

Answers

(a) The inputs are by snowfalls and avalanches. The outputs are by evaporation, deflation and meltwater.

(b) The global climate related to latitude is the overall spatial variation. Climatic change may alter this. At the valley glacier scale, in lower latitudes, altitude and

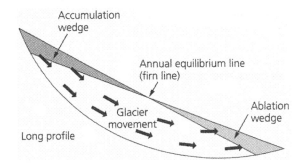

Figure 14.11 *Basis of glacier mass balance. In a glacier that is in equilibrium the ablation wedge equals the accumulation zone. The accumulation zone exceeds the ablation wedge in an expanding glacier, and is less than it in a retreating glacier.* After Figure 6.5 in Clowes, A. and P. Comfort (1982) Process and landform. *Oliver & Boyd, Edinburgh.*

aspect are important. Aspect may promote increased accumulation on north-facing slopes (northern hemisphere), or on the lee slopes of prevailing wind direction due to orographic eddying effects. As the snowline is higher in lower latitudes only high peaks will be able to provide an input of snow.

(c) Mass balance will vary over different time-scales. Short-term climatic change in either temperature or precipitation will produce a relatively fast response in smaller-scale systems, e.g. valley glacier, whilst the same variables will take longer to influence less responsive larger ice masses, e.g. ice caps. Longer-term changes reflect longer climatic cycles such as in relation to the Earth's orbital position.

2   Figure 14.20 shows ice contact depositional features. Answer the following question.
   (a) Compare and contrast the depositional forms shown on the diagram in relation to:
     i   environment of deposition;
     ii material structure and composition.

Answers

(a) Eskers are deposited below or within the ice, kames on the ice and kame terraces at the ice margin. Both eskers and kames are deposited in a high energy environment close to the ice front. The material producing kame terraces is deposited in a lower-energy environment. The energy in the environment is related to the flow of water.

(b) As the deposition is water-related the material is generally well sorted, rounded and initially stratified. The disturbance of stratification is related to the relative fall due to gravity once the ice melts. Deposits higher up in the glacier will be disturbed more due to the greater vertical fall.

Short-answer questions

1   What is the snowline and how does it vary?

Answer

The snowline is the lower edge of permanent snow in high latitudes or mountainous regions. It varies with latitude, altitude and aspect and generally drops from about 5,200 metres at the equator to sea level at the poles.

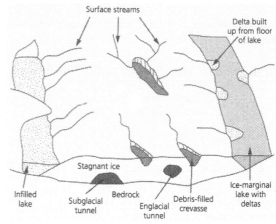

a Glacial landscape

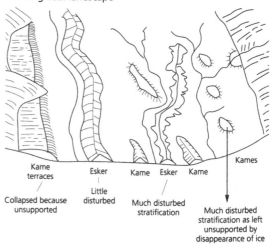

b Post-glacial landscape

Figure 14.20 *Ice-contact depositional features. Characteristic ice-contact features include kames and eskers (see text for explanation). After Figure 7.12 in Clowes, A. and P. Comfort (1982)* Process and landform. *Oliver & Boyd, Edinburgh*

2   How do the dominant cold air masses affect climate in the tundra zone?

Answer

The cold air masses suppress summer temperatures and produce very low winter temperatures. Little water vapour is held in the cold, descending air so annual precipitation is limited, usually less than 250 mm a year.

3 What values does the Antarctic have that makes it attractive for conservation?

Answer

The Antarctic represents one of the few almost natural environments on Earth. It possesses a unique habitat and its sea areas are extremely productive, maintaining a large biotic community. The scale and position of the Antarctic means it plays a key role in regulating global climate, and in monitoring future climatic change.

4 How do glacier types vary?

Answer

Glacier types vary both in size and location. Alpine glaciers are confined to valleys and are found in high mountain areas. Piedmont glaciers are extensive ice sheets formed by the coalescence of a number of valley glaciers at the foot of mountain ranges. Ice-cap glaciers comprise a large, central ice sheet-covered plateau surrounded by a series of valley glaciers flowing down the plateau sides. Continental glaciers are the most extensive, covering entire land surfaces such as Antarctica.

5 What is the process of accumulation?

Answer

Snow falling above the snowline accumulates to great depths as precipitation exceeds melt. Individual snowflakes pack together under their own weight to form granular snow (firn) where the individual grains become united to form a cohesive structure.

6 In what ways does glacial ice move?

Answer

Glacial ice moves either by basal sliding on a film of meltwater between the ice and the ground surface or by internal plastic deformation of the compacted ice (laminar flow) in the main body of the glacier.

7 Describe how global warming may affect glacial mass balance.

Answer

This may be a significant problem as it might promote climatic positive feedback at the regional level. A rise in temperature will increase the rate of ablation but this may be offset by an increase in snowfall as temperate depressional weather regimes are pushed polewards. The predicted global model is that ablation will dominate, especially in the southern hemisphere, with melting of the Antarctic peninsula causing a 1mm rise in sea level within 40 years.

8 What is abrasion?

Answer

Abrasion is an erosional process whereby entrained rock fragments in the base of a moving ice mass scour the surface of the ground. This may result in the smoothing of rough surfaces, particularly when the rock fragments are very fine (rock flour), and the scratching of smooth surfaces to produce striations.

9 List the ways glaciers transport sediment.

Answer

Glaciers move sediment in a number of ways: some is frozen to the base of the ice; some is frozen within the ice; sediment may be transported on the surface either solely on the ice or at the ice margin; and some sediment is pushed along ahead of the ice as it advances.

10 Outline the locations of ice melt from a glacier.

Answer

Ice melting or ablation may occur due to frictional drag on the underlying rock, causing basal melt. Basal melt might also be the product of pressure of overlying ice; this may also occur within the ice. Finally, melting occurs on the surface and at the sides of the ice due to higher temperatures in the surrounding environment.

Additional references

Vaughan, D. G. and Doake, S. C. M. (1996) 'Recent atmospheric warming and retreat of ice shelves on the Antarctic peninsula'. *Nature* 379 (6563), 328–31.

Antarctic air temperature trends illustrate regional climate warming and the break up of ice shelves. The spatial extent of Antarctic ice is seen as closely influenced by climate change. A temperature threshold for the ice-shelf system suggests applications supporting these as critical indicators of climate change.

Web site

www-nsidc.colorado.edu

This is the web site of the United States National Snow and Ice Data Center. It contains a very large data archive of relevant resources.

Aims

- To examine the structure and operation of ocean systems.

- To emphasize the dynamic interactions between land and sea at the coastal boundary.

- To evaluate the importance of the coastal system.

Key-point summary

- The oceans not only comprise the aquatic environment but are dynamically linked to the lithosphere in relation to ocean crustal genesis and ocean floor topography.

- The sheer spatial scale of the oceans and their adjacency to the majority of the world's human population who inhabit the coastal zone make them extremely important. Due to this proximity to large human populations they are exploited as a natural resource, but also suffer degradation from incompatible and unsustainable use. The results of this exploitation may lead to system instability. However, compared to terrestrial systems, scientific knowledge is limited and outcomes uncertain. Similarly, global ownership of the oceans beyond coastal political limits means policy making needs to be of an international dimension.

- At a large scale the oceans can be spatially divided into *basins*. These may be further subdivided into areas of *continental shelf*, *continental slope*, *continental rise* and *ocean floor*. This classification reflects general underwater topography and therefore water depth. These variables affect *temperature*, *light* and *nutrient* status of the water with consequent effects on plant and animal life.

- The main properties of ocean water reflect links to other major systems. *Salinity* relates to continental denudation systems; *temperature* broadly reflects latitudinal climatic conditions and *wave* characteristics are strongly influenced by wind.

- The ocean system must be viewed as three-dimensional, with *ocean currents* producing flows through the system comparable to global wind circulation, and the influences acting on it. These flows influence *surface weather* conditions and *biological productivity*.

- Ocean currents suffer periodic shifts in both time and space, reflecting the influence of other environmental systems. The El Niño current is an extreme example of this. Scientific knowledge of the processes and causes are limited to temporal correlation with other natural systems' disturbance, e.g. seismic activity in the lithosphere and *ITCZ* movement in the atmosphere. Correlations exist with the effects of El Niño on other systems. These cover a wide range of spatial scales and locations.

- *Tides* illustrate the influence of a much larger-scale system on the oceans. The Sun and Moon exert gravitational attraction on bodies of water, effectively influencing the surface water level. Within this general movement smaller-scale flows (*tidal currents*) reflect the influence of coastal relief features.

- The oceans are an important *global common* and as such are under threat, to a large extent, from over-exploitation and sustainable use.

- The oceans are amongst the most productive ecosystems on Earth. This productivity is spatially defined by sunlight penetration and is therefore principally confined to the upper parts of the ocean. The resources are thus easily exploitable. To remain sustainable, by population replacement, the *maximum sustainable yield* of fish catch should not be exceeded. The balance between yield and replacement in a sustainable system is a good example of *steady-state equilibrium* in an open system. Overfishing causes a breakdown of this equilibrium, with recovery times for individual species being lengthy.

- Non-biotic resources, minerals and energy, unlike fish are far harder to exploit. Minerals on the ocean floor are less accessible. Renewable energy from the oceans is exploited to a limited degree but this process is costly to establish and for viability needs to be on or near the coast. This causes conflict with coastal ecosystems.

- The ocean system is heavily influenced by pollution generated from the land. The concentration of the world's population on or near the coast makes this inevitable. Pollution causes spatial problems in two ways. Local coastal conditions may concentrate the pollution in particular areas, e.g. eutrophication, and the dispersing nature of ocean currents spreads point source pollution vast distances to affect other areas, even though concentrations are usually lowered by dilution. *Oil pollution* from tankers might be considered as internal to the use of the oceanic system (by transport) but irrespective of whether sources are internal or external, the oceans are considered as 'global commons' which means that global responsibility and management are needed to combat the problem. Again, sovereignty tends to override the common good, though there have been efforts to work together at the scientific level (*ecosystem units*) and at the regional scale by coastal nations producing co-operative management agendas, e.g. *Mediterranean Action Plan*.

- Coasts are highly dynamic open systems at the interface between land and sea. As they respond quickly to change in either of these two major environmental systems and as they are adjacent to large populations their management is of great importance.

- Coastal landforms reflect the processes of erosion and deposition. The geological control influences the rates of both these processes via resistance and material supply whilst marine water movement powers the processes within the influence of the marine environment. Beyond this, atmospheric processes operate, e.g. beyond the high-water marks.

- As part of the coastline, *estuaries* represent the junction between the marine component of the hydrosphere and the freshwater land-based riverine component. Their varied inputs and morphology make them extremely productive and biologically valuable ecosystems.

- Because of its spatial proximity and highly dynamic nature the management of processes at the coast is extremely important to humans. Threats to the occupied coastal zone occur at a variety of scales. Globally the threat of sea-level rise as a consequence of global warming is seen as the major hazard to coastal areas. Management options vary from the traditional *hard engineering* approach to a more sympathetic approach utilizing the potential of natural systems (*soft protection*).

- *Beaches* dissipate wave energy and are constantly reacting to a variety of inputs to achieve a *dynamic equilibrium*. Where sediment supply is available they provide excellent coastal protection, including the formation of dune systems by the accumulation of wind-blown sand as an extension inland. Threats to beach and dune system sea-defence roles occur due to a restriction in sediment supply. This is usually the result of sediment extraction by humans exploiting the resource or natural processes of redistribution, e.g. longshore drift. The key management technique to maintain beach and dune systems is to *stabilize* the flow through the system and ensure adequate input by *replenishment*.

- Coastal zones face a wide range of natural hazards. The two most difficult to manage are storm surges and coastal flooding. Both represent direct links with other systems. Storm surges require strong winds (weather system) to cause already high water levels to overtop defences; coastal flooding is also caused by tsunamis (tectonic system) and tidal inundation (cosmic system).

- The most complex potential coastal flood hazard is sea-level change. This is affected by both natural processes (climate change and relative earth–land movement) and human-induced change (enhanced global warming). Coastal flooding due to sea-level rise is a good example of the truly interactive operation of the Earth system, with self-regulation going beyond the interests of a single species – humans.

Main learning hurdles

Oceanic movement and flows

Students with a non-scientific background may have problems regarding the relative motion of water within waves. The instructor should use Figure 15.4 to ensure students understand wave motion and the relationship to the wind system. A revision of gravitational force and atmospheric circulation (Chapters 3 and 8) is useful when considering currents and tides.

Key terms

Abyssal zone; assimilative capacity; atolls; bathyal zone; beach; Braer; cliffs; continental shelves; coral reefs; dilution; dunes; Earth Summit; El Niño; *Exxon Valdez*; estuaries; eustatic change; fishing; flood and ebb currents; flooding; global ocean commons; Gulf Stream; gyres; hadal zone; harvesting; isostatic change; Law of the Sea Convention; longshore drift; mariculture; marine eutrophication; neritic zone; North Atlantic Drift; ocean currents; ocean floor; phytoplankton; productivity; red tides; rias; salinity; sea-level change; storm surges; sustainable yield; temperature; tides; thermal energy; wave energy.

Issues for group discussion

Discuss the likely effects of global warming on the world's coastal flood hazard

The students should review the breadth of both sea-level rise and the flood hazard on coasts by reading Tooley (1989) and Spencer and French (1993). Students should then be able to emphasize the natural spatial variation, and this can be developed in terms of human management by reading the case studies of Connell (1990) and Younger (1990). The final emphasis should focus on the unequal abilities between the economically developed and less developed worlds in coping with the problem.

Discuss the view that we are overexploiting the oceanic resource

Gwyer (1991) produces a well-balanced article concerning the use of the fish resource. Being critical is the key focus in this discussion. Students should question whether we are overfishing and how do we know? The article by McKie (1989) should introduce the students to other marine resources we hardly exploit. What would be the impacts of this (both positive and negative) and would exploitation compromise other marine resources? This would

provide an interesting end debate. The instructor should encourage the students to agree one way or another. This will probably indicate the difficulty of agreeing policy for global commons.

Selected reading

Connell, J. (1990) 'The Carteret Islands: Precedents of the greenhouse effect'. *Geography* 75 (2), 152–4.
A brief case study of the effects of global warming, via sea-level rise, on susceptible island coasts. System links are clearly implicated with a variety of identified hazard outcomes.

Gwyer, D. (1991) 'Fishing for trouble'. *Geographical Magazine* 63 (7), 20–3.
The economic and conservation aspects of overfishing are examined. Highlights the strength of the economic argument and to a certain extent the limits of scientific knowledge of the marine environment.

Marcus, W. A. and Kearney, M. S. (1991) 'Upland and coastal sediment sources in a Chesapeake Bay estuary'. *Annals of the Association of American Geographers* 81 (3), 408–24.
An investigation of the coastal sedimentation and erosional processes. The dynamism of the land–sea interface is depicted in flows between erosive areas of the coast and estuarine sediment stores. There is an assessment of flows via river systems from terrestrial systems to provide an integrated overview.

McGregor, G. R. (1995) 'The tropical cyclone hazard over the South China Sea 1970–1989: Annual spatial and temporal characteristics'. *Applied Geography* 15 (1), 35–52.
The South China Sea cyclone hazard is illustrated and linked to larger spatial atmospheric disruption by El Niño–Southern Oscillation events. Interactions between the oceanic system and the atmosphere are demonstrated through time. Useful applications are suggested in relation to indicating the spatial potential of the cyclone hazard.

McKie, R. (1989) 'Mining the depths'. *Geographical Magazine* 61 (7), 16–20.
A non-technical account of the oceanic mineral resource. Usefully outlines the claims of ownership on the ocean commons.

Penney, T. R. and Bharathan, D. (1987) 'Power from the sea'. *Scientific American* 256 (1), 74–80.

Ocean thermal energy conversion is explained. This is contextualized in relation to its future use in the global energy mix. Potential areas of exploitation in the ocean system are spatially identified in relation to temperature ranges and sea-floor relief.

Spencer, T. and French, J. (1993) 'Coastal flooding: Transient and permanent'. *Geography* 78 (2), 179–82.
A short article illustrating the scale and various responses to coastal flooding. Level of economic development is seen as the major factor related to the scale of disaster. Sea-level change trends put the potential hazard of coastal flooding in context.

Tooley, M. (1989) 'The flood behind the embankment'. *Geographical Magazine* 61 (11), 32–6.
This article discuss the dynamism of the British coastline in both time and space. The influence of the global climatic system on sea-level change is usefully discussed. Good illustrations depict coastal outline change in relation to the worst case scenarios of sea-level rise resulting from global warming.

Younger, M. (1990) 'Will the sea always win? Coastal management in north-east Norfolk'. *Geography Review* 3 (5), 2–6.
A case study of the various system components interacting to produce differential coastal erosion. The system analogy is used throughout with a consideration of the range of management options.

Textbooks

Bergesen, H. and Parmann, G. (eds) (1993) *Green Globe Yearbook*. Oxford University Press: Oxford.
This edition contains a key section on the pollution status of the marine ecosystems in the 1990s.

Carson, R. (1997) *The Sea Around Us*. Oxford University Press: Oxford.
The classic work brought up to date with an additional chapter by Jeffrey Levinton. The ecological nature of the oceanic component of the hydrosphere and its importance are systematically evaluated.

Clark, R. B. (1992) *Marine Pollution*, 3rd edn. Oxford University Press: Oxford.
A readable introductory text that explains the marine system and the threats to it from human polluting activities.

Davis, R. A. Jr (1997) *The Evolving Coast*. W H Freeman: Basingstoke.

A clear introductory text describing the variety of scales of influence on the formation and character of coasts. Large-scale long-term influences such as crustal plate movement and sea-level change are considered alongside short-term processes such as waves, tides and weather. Very readable and well illustrated.

Ince, M. (1990) *The Rising Seas*. Earthscan: London.
This text provides a good overview of sea-level change. It is written in a student-friendly style with a focus on marine inundation illustrated with a range of case studies.

McGoodwin, J. R. (1992) *Crisis in the World's Fisheries: People, Problems and Politics*. Stanford University Press: Stanford.
A comprehensive investigation of the marine fishery resources. Outlines the variety of resource demands and uses. Contains a very useful appraisal of the application of environmental law as a policy tool.

Essay questions

1 Discuss the processes that may result in estuarine salt marshes changing from areas of overall accretion to zones of rapid erosion.
2 Examine the key factors in the development of coastal sand dune systems.
3 Examine the significance of waves in depositional processes operating in the coastal environment.
4 Explain the main ways in which the oceans interact with other major environmental systems.
5 Critically evaluate the role of evidence of sea-level change during the Holocene in predicted future sea-level change.
6 What are the alternatives to 'hard' engineering solutions used in coastal management.
7 Discuss the impact for the marine environment of oil spillage and subsequent clean-up operations.
8 Examine how far examples of recent coastal defences illustrate an increasing awareness and understanding of geomorphological processes.
9 Outline the major issues threatening the coastal zone using examples from around the world.
10 Evaluate the problems of managing the 'global ocean commons'.

Multiple-choice questions

Choose the best answer for each of the following questions.

1 The percentage of the world's population living in coastal areas by the end of the century is expected to be:
   (a) 45%
   (b) 50%
   (c) 60%
   (d) 70%   *

2 How much of the total world fish catch is supplied by coastal waters?
   (a) 35%
   (b) 55%
   (c) 95%   *
   (d) 98%

3 The bulk of the dissolved material in sea water is:
   (a) calcium
   (b) potassium
   (c) sodium   *
   (d) magnesium

4 The average temperature of sea water at the poles is:
   (a) −32° centigrade
   (b) −1.4° centigrade   *
   (c) 0° centigrade
   (d) 4° centigrade

5 The reversal between flood and ebb currents when water is relatively calm is known as:
   (a) heavy water
   (b) a neap tide
   (c) a spring tide
   (d) slack water   *

6 Biological productivity in the oceans depends on the primary producers, which are:
   (a) phytoplankton   *
   (b) zooplankton
   (c) krill
   (d) seaweeds

7 Which of the following contributed most to the world's increase in fish catches between 1980 and 1986:
   (a) Japan
   (b) USSR
   (c) Australia
   (d) Latin America   *

8 The maximum sustainable yield (MSY) reflects:
   (a) population growth
   (b) depleted stocks
   (c) carrying capacity   *
   (d) total stock

9 Optimum sustainable yield (OSY) is usually calculated as:
   (a) 10% of MSY
   (b) 25% of MSY
   (c) 50% of MSY   *
   (d) 75% of MSY

10 In the North Atlantic overfishing has seriously depleted populations of:
   (a) plaice and sole
   (b) dogfish and monkfish
   (c) cod and haddock   *
   (d) mackerel and herring

11 The most use of the ocean thermal energy conversion (OTEC) has been made in the:
   (a) Mediterranean sea
   (b) Arctic Ocean
   (c) North sea
   (d) Pacific Ocean   *

12 The majority of marine pollution monitoring is based in:
   (a) marine trenches
   (b) deep ocean sediments
   (c) the Arctic
   (d) the coastal zone   *

13 Between 1975 and 1978 how many oil slicks were reported?
   (a) 100,000   *
   (b) 10,000
   (c) 1,000
   (d) 100

14 What is the first stage in China's coastal zone management strategy:
   (a) engineering works
   (b) an inventory   *
   (c) selective development
   (d) legislation

15 A low ridge of sand projected from the shore out to sea is called a:
   (a) spit   *
   (b) tombolo
   (c) bar
   (d) haff

16 NEP stands for:
   (a) National Energy Program
   (b) Natural Environment Program
   (c) Natural Ecosystem Protection
   (d) National Estuary Program   *

17 At the beginning of the Holocene, about 11,000 years ago, what was the position of global sea level?

    (a) 54 centimetres higher
    (b) 54 centimetres lower
    (c) 54 metres higher
    (d) 54 metres lower   *

18 Land-level changes affecting relative sea level are known as:

    (a) exostatic
    (b) isostatic   *
    (c) eustatic
    (d) endostatic

Figure questions

1   Figure 15.6 shows the distribution of the main surface ocean currents. Answer the following questions.

    (a) What drives the ocean current circulation?
    (b) How does ocean circulation influence global climate?
    (c) Why is it important?

Answers

(a) On the surface the ocean currents are driven by prevailing winds, and modified directionally by the Coriolis force. However, their circulation at depth within the ocean is related to density gradients reflecting temperature and salinity of the water masses.

(b) The ocean current system influences global climate by redistributing vast amounts of heat energy around the Earth. This interacts with the lower atmospheric system.

(c) This global circulation of oceanic heat is important in defining the pattern of climate. Any

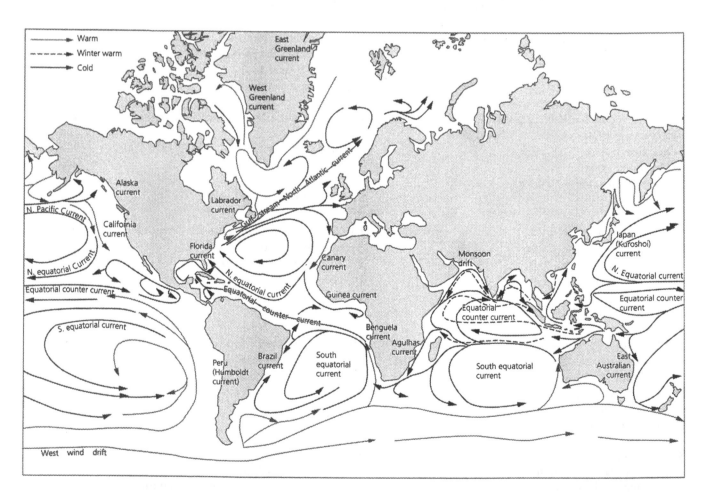

Figure 15.6 *Distribution of the main surface ocean currents. See text for explanation. After Figure 6.1 in Doerr, A. H., (1990)* Fundamentals of physical geography. *Wm.C. Brown Publishers, Dubuque*

change in the ocean circulation system would have profound climatic effects and help to drive rapid climate changes. If the currents did not operate, heat from the equator would not be moved polewards, causing high-latitude cooling and possibly global cooling via an onset of glacial activity in these regions.

2 Figure 15.9 shows the zonation of the coast. Answer the following questions.
  (a) What is the sum of these zones known as?
  (b) Describe the interaction of waves in these zones.
  (c) What sediment is found in the low water breakers area?

Answers

(a) The total area is known as the littoral zone.

(b) In the offshore zone incoming waves start to interact with the sea floor. This modifies the waves and they break and enter the shore zone. At low tide they stay below the beach but at subsequent higher tides they run up the foreshore, dissipating energy by friction with the rising slope. The waves become an increasingly shallow film of water towards the high-tide mark. The backshore is only infrequently subject to severe storm waves when weather conditions and high tides prevail.

(c) The low water breaker zone is a turbulent high-energy environment able to entrain and carry landward all but the coarsest material. This leaves the coarser material in situ, comprising the sediment in this zone.

Short-answer questions

1 List the key areas where greater scientific understanding of the oceans is needed.

Answer

A great deal is still uncertain about the ocean's role in other systems' operations and processes. These include: environmental hazards, e.g. tsunamis; climatic variation, e.g. the El Niño phenomenon; climatic change, e.g. global warming; ecosystem productivity; biochemical stores and global system regulation.

2 Describe the major influences on wave size.

Answer

Wave size depends on wind speed, with increasingly strong winds producing bigger waves. The longer the duration of winds the larger the waves. Wave size also increases with fetch distance and may be influenced by contact with waves generated elsewhere.

3 What is El Niño?

Answer

This is an intermittent reduction in cold water upwelling off the coasts of Peru, Ecuador and Chile. The net effect is warmer water off these coasts.

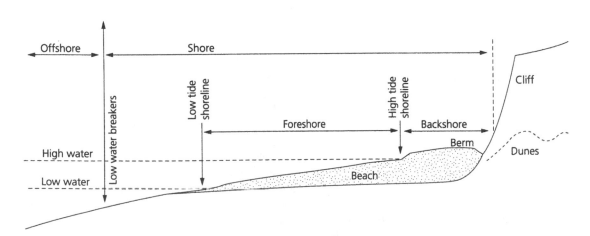

Figure 15.9 *Zonation of the coast. The coast can be sub-divided into a number of distinct zones on the basis of water levels.*
After Figure 9.2 in Clowes, A. and P. Comfort (1982) Process and landform. *Oliver & Boyd, Edinburgh*

4 List the factors that affect fish yields.

Answer

A number of factors affect fish yield: technology where modern boats and techniques take larger catches; economic demand driven by cultural taste; and previous resource exploitation which will reduce yields if overfishing has taken place.

5 Describe the options available for exploiting renewable ocean energy resources.

Answer

The main options are exploiting the ocean thermal gradient to provide energy and exploitation of tidal power and wave energy. Possible future options involve: tapping submarine geothermal sources; using marine biogas; exploiting the thermo-haline gradients; and constructing offshore wind forms.

6 What was the *Exxon Valdez* disaster?

Answer

This was a massive oil spill off the coast of South Alaska when the supertanker *Exxon Valdez* ran aground in 1989. Apart from the scale of the slick, covering 12,400 square kilometres, it was the sensitivity of the local ecosystems to the spill that exacerbated the ecological disaster.

7 What is a 'red tide'?

Answer

A red tide is the change in the colour of sea water due to severe algal blooms.

8 Describe how you might control global ocean pollution?

Answer

Controlling ocean pollution can be achieved by reducing its input from source areas, e.g. by industrial waste management. This will require a global commitment at the international level but will also require local awareness due to the concentration of people living in the coastal zone. Pollution from ocean transport requires reducing the shipping hazard (navigation systems) and reducing bad practice (tank flushing at sea).

9 What is a tombolo?

Answer

A tombolo is a narrow sand or shingle bar connecting an island to the mainland or with another island.

10 Suggest two means of protecting the coast.

Answer

The coast may be protected by hard engineering schemes such as sea walls. A softer option is to promote natural processes such as beach development to absorb the wave energy.

Additional references

Bray, M. J. and Hooke, H. H. (1996) 'Managing the Wessex coast: Towards an integrated approach'. *Geography Review* 10 (1), 31–3.
A clear illustration of the spatial distribution of erosion and sediment transport along the Wessex coast is presented. Knowledge of processes is used to illustrate the effectiveness of an holistic approach to managing the coastal environment.

Broecker, W. S. (1995) 'Chaotic climate'. *Scientific American* 273 (5), 44–50.
Links the global current circulation with climate in explaining climatic change. This is viewed as a critical element in determining rapid change in weather patterns. Possibilities of future predictions are also discussed.

Case, R. (1997) 'Managing the Wessex coast: Christchurch Bay'. *Geography Review* 10 (3), 20–5.
An interesting case study of human intervention in the geomorphological cycle of coastal development. System processes are well illustrated and a multidisciplinary, integrated management strategy emphasized.

Davis, B. (1996) 'Oracle of the oceans'. *New Scientist* 152 (2051), 27–31.

Introduction to a monitoring system of marine sensors providing information on the world's oceans. Links many applications to the increased knowledge of how the ocean system operates. Good examples relating to forecasting El Niño events and biological activity are discussed.

Iannotta, B. (1996) 'Mystery of the Everglades'. *New Scientist* 152 (2055), 34–7.
Links the land agricultural system to ocean resources via coastal pollution. Chemical and biological linkages are discussed in a case study of Southern Florida.

MacDonald, A. M. and Wunsch, C. (1996) 'An estimate of global ocean circulation and heat fluxes'. *Nature* 382 (6590), 436–9.
Illustrations depict global ocean circulation patterns. Exchange of heat and gases between the oceans and the atmosphere is driven by these patterns with direct linkages to the general global climate system.

Pearce, F. (1996) 'Crumbling away'. *New Scientist* 152 (2061), 14–5.
Focuses on dredging operations off the Norfolk coast. Useful case study of how human action has upset the balance of the coastal erosion/deposition system.

Safina, C. (1995) 'The world's imperiled fish'. *Scientific American* 273 (5), 30–7.

A regional look at the combined threats on marine fish resources. The exploitation of this limited resource is assessed in relation to future sustainability with management options outlined.

Schneider, D. (1997) 'The rising seas'. *Scientific American* 276 (3), 96–101.
A well-illustrated article examining the various sea-level rise predictions associated with global warming. Spatial extent and rate of change are discussed in the light of these predictions. Possible disasters are linked to the influence of local weather.

Spurgeon, D. (1997) 'Fisheries science: All at sea when it comes to politics?' *Nature* 386 (6621), 105–7.
A short overview of the problem of overfishing. Clearly states the global nature of the problem and discusses the use of scientific evidence in policy formation.

Web site

www.mth.uea.ac.uk/ocean/oceanography
A well-indexed site with updated links to a variety of marine orientated sites of a multidisciplinary nature. Usefully incorporates its own search engine.

Aims

- To examine the components of the biosphere.

- To explain the significance of biodiversity.

- To use the theories, principles and approaches of ecology to understand evolution, species extinction and conservation practices.

Key-point summary

- The *biosphere* is the major natural system that supports life on Earth. Most living organisms (*biota*) are found on or near the surface of the Earth. Therefore, the biosphere may be defined as a comparatively thin layer comprising a lower boundary within the lithosphere and an upper boundary within the atmosphere. People may be viewed as the most important component of the biosphere due to their conscious ability to alter the system's components and processes. The scale of human impact has increased through time from the local and regional to produce changes of global significance. This interference with our life-support system means critical sustainable thresholds are being approached.

- Assessment of human action on the biosphere requires knowledge of the system. *Ecology* studies the interactions between all plants, animals and their environments. Study of these relationships allows an insight into the structure and function of nature and how it responds to change. The great breadth of variables within the system (*biotic* and *abiotic*) has fostered a multidisciplinary approach; with the scale of the effects and potential system response encouraging international research effort.

- *Biodiversity* reflects the richness and variety of nature's species. These biotic components are classified into three groupings – *plants*, *animals* and *protista*.

- Our knowledge of the scale and pattern of biodiversity is limited. Estimates of the total number of species vary between 5 million and 80 million. As the system inventory is so uncertain there is much speculation regarding the outcomes of the human impact at the global scale. More certainty exists at the local scale where management outcomes can be observed more readily.

- The conservation of the Earth's biodiversity is essential for a variety of reasons. Ethically every species has the right to exist. Culturally the diversity of nature enriches our lives. And economically the wide range of genetic material can be used in agriculture, medicine and industry. Such a range of reasons promotes a multi-disciplinary approach to sustainable usage.

- Species have disappeared naturally through time due to changing environmental conditions. However, the acceleration of *habitat* destruction by humans has put many more species at risk over a much shorter time-scale. This has interfered with the natural plant and animal capability of evolving to longer-term environmental change.

- *Darwin*'s theory of natural selection describes how the biotic system operates to provide species best adapted to particular environmental conditions. Thus, distribution patterns of communities closely match the spatial mosaic of habitats on the Earth's surface. These large-scale patterns may be altered by significant human interference such as global warming.

- Evolutionary principles may be applied through *genetic engineering*. This produces species more adapted to survival, e.g. pest/disease resistance, offering benefits of better resource use for society.

- The evolutionary process that produces genetic diversity is important as it underpins the biosphere by producing organisms that helped create the atmosphere. Thus, there is a *symbiotic* relationship between biota and biosphere. The climatic system naturally impacts back, through its longer-term cycles, by producing conditions suitable for domination by particular groups of plants and animals. These groups are used to define the geological time periods (eras) of the Palaeozoic, Mesozoic and Cenozoic.

- Counter-balancing the evolution of species (*speciation*) is *extinction*. This balances the Earth's biodiversity. Naturally through time this balance remains relatively constant, though occasionally mass extinctions occur due to rapid global-scale system change, e.g. the catastrophic atmospheric effects of a meteorite strike. Human activities either directly (hunting) or indirectly (habitat clearance) unbalance this natural equilibrium by accelerating extinction.

- *Conservation* is designed to reduce this imbalance.

- Conservation requires a multidisciplinary approach. It is a dynamic concept, distinguishable from *preservation*. Good conservation management relies heavily on habitat knowledge and the spatial awareness of how they interact at various scales.

- In practice many methods may be employed in conservation management. Often the management has to be viewed in the light of other competing uses, but it usually results in spatial decisions, either protecting areas *in situ* or setting aside areas *ex situ* if the local policy decisions preclude effective conservation of the natural habitat and its attendant species. The second option is less preferable.

- *Protected areas* have a variety of designations and consideration of the use of the space. However, as a designated area, conservation is treated as a priority. Spatial boundaries defining protected areas allow applied scientific work to complement and enhance our knowledge of ecosystem dynamics. This is usually accomplished by monitoring biotic interactions. The use of protected area status recognizes the importance of scale in preserving system integrity and inter-action. At the small scale *nature reserves* may involve relatively little interaction. At the meso-scale *national parks* require the consideration of a number of human–physical interactions and their management requires broader objectives. Yet their larger size promotes a more stable natural system if the level of protection is high. At the largest scale, *biosphere reserves* acknowledge major environmental system interaction and the need to conserve the balance. Yet these face the greatest conflicts, whilst potentially yielding the greatest benefits, from regional development objectives.

- Sometimes spatial habitat protection is not enough to protect individual species that have an economic worth to human society. The effective introduction of an unnatural predator (humans) into the system means special measures need to be taken either in situ (e.g. armed wardens) or ex situ by captive breeding programmes.

Main learning burdles

Terminology of protection

Students often get confused between the definitions of conservation and preservation. This is often reinforced by the interchangeable use of the terms in the popular media. Reference to Box 16.31 should be emphasized by the instructor and a consideration of other non-biotic examples would be useful, e.g. historic buildings, cultural practices, etc.

Key terms

Adaptation; biosphere reserves; biota; biotechnology; classification; conservation; Darwin; ecology; euphotic zone; evolution; extinction; 'gap analysis'; genetic engineering; intergenerational equity; life-support system; Mesozoic; national parks; natural selection; nature reserves; ontogeny; Palaeozoic; Phanerozoic; phylogeny; preservation; protected areas; transgenic; wilderness.

Issues for group discussion

Discuss the contention that nature conservation should take priority over resource development

This discussion should examine the range of activities that may take place in an area. Hirsch (1994) takes a wide view of this. Students should develop their own ideas as to what they expect from natural areas and refer to the various levels of conservation priority outlined in the chapter and assess how this might compromise their expectations. National parks and wilderness comparisons should be encouraged by the instructor. Wider development issues regarding global regions and their resources should eventually be considered.

Discuss whether life is a resource

After reading Simmons (1988) the students should discuss the importance of individual species. The instructor should control the discussion as student views may become polarized on this issue. The discussion should develop into a consideration of species communities and their habitats. Wilson (1989) evaluates this as well as considering the range of threats. Broad management and individual responsibility issues may summarize the discussion. A good thought to leave the discussion with is: how does my normal consumption pattern of, for example, consumer goods affect other species?

Selected reading

Hirsch, P. (1994) 'Protected areas and people'. *Geography Review* 8 (2), 37–41.
Considers the spatial design of protected areas. This adopts an holistic viewpoint, arguing for integrated management between human systems and natural resources.

Simmons, I. (1988) 'Life as a resource'. *Geography Review* 1 (3), 32–7.
An outline discussion of our use of the biotic resources of the Earth. Links into changes in the ecological systems and describes system outcomes via distribution and spatial patterning of these resources. Contains a useful glossary. NB: This is part two of a three-part series of articles in consecutive issues.

Wilson, E. O. (1989) 'Threats to biodiversity'. *Scientific American* 261 (3), 60–9.
Quantifies the rate of habitat destruction in the tropics with the rate of species extinction. Clear system links between the scale of habitat loss and the increased degradation of biodiversity are explained. Consequences of this trend of reduced biodiversity are suggested.

Textbooks

Adams, W. (1986) *Nature's Place: Conservation Sites and Countryside Change*. Allen and Unwin: London.
A useful discussion of the legislation designating and protecting particular spatial areas. The effectiveness of this in relation to nature conservation is evaluated.

Adams, W. M. (1996) *Future Nature: A Vision for Conservation*. Earthscan: London.
Outlines the philosophies behind nature conservation. Provides a broad view combining both habitat and landscape in an holistic view.

Barbier, E. B. (1990) *Elephants, Economies and Ivory*. Earthscan: London.
Puts forward the case for conservation and sustainable resource management. Good, interesting case studies provide interesting debate between a range of economic interests and environmental considerations.

Evans, D. E. (1996) *A History of Nature Conservation in Britain*, 2nd edn. Routledge: London.
Considers a variety of conservation practices and policy based on pressure from conservation organizations. The change in emphasis of what conservation means and how it might be attained, through time, is convincingly explained with reference to examples.

Huggett, R. J. (1997) *Environmental Change: The Evolving Ecosphere*. Routledge: London.
Change through time of the environment is the main focus of this introductory text. The interdependent nature of the Earth's major environmental systems is linked to a wider consideration of the solar system's inputs.

Jeffries, M. (1997) *Biodiversity and Conservation*. Routledge: London.
This text provides an introduction to a wide range of themes from a variety of disciplines. Worldwide examples are used to explain core topics such as species extinction. Biosphere components are examined to set up a rationale for conservation based on integrating protection and resource use.

Jones, A. M. (1997) *Environmental Biology*. Routledge: London.
An introductory text linking scientific biological systems' understanding to the holistic operation of the biosphere. Patterns and processes of terrestrial

and aquatic ecosystems are presented clearly with a progressive discussion through to more complex concepts.

Kellert, S. R. (1997) *The Value of Life: Biological Diversity and Human Society.* Earthscan: London.
A wide-ranging evaluation of the values of biodiversity. It examines the living world (including humans) as an integrated whole and evaluates species destruction as threatening both our physical and cultural worlds.

Morgan, S. (1995) *Ecology and the Environment.* Oxford University Press: Oxford.
This book brings together the interactions between physical and biological factors with a major emphasis on biogeochemical cycling. The various mixes of the components are explained as underpinning the Earth's distribution of biomes. Within the biome scale individual food chains and webs are illustrated.

Simon, N. (1995) *Nature in Danger: Threatened Habitats and Species.* Oxford University Press: Oxford.
A global perspective is offered on the range of threats to endangered species and their environments. Over-exploitation is seen as the key issue in the degradation of ecological resources and mechanisms for protection are expounded.

Szaro, R. C. and Johnston, D. W. (1996) *Biodiversity in Managed Landscapes: Theory and Practice.* Oxford University Press: Oxford.
Focuses on the need for comprehensive scientific data as the foundation for rational decision making. Conservation policy is the key aim of the work and useful case studies, illustrating a variety of spatial scales, are documented. The scientific basis needed for sustainable policy decisions is emphasized.

Essay questions

1 Discuss how the introduction of exotic species into an ecological system will affect the system's balance.
2 Discuss the importance of continental movements on the distribution of flora and fauna.
3 Discuss the advantages and disadvantages of genetic engineering with particular reference to biodiversity.
4 Evaluate Darwin's theory of natural selection.
5 Compare and contrast the disciplines of biogeography and ecology.

6 Given the problems of economically developing countries, how do you justify conservation as a priority?
7 With reference to ecosystem functions, suggest factors that you consider are relevant when establishing nature conservation areas.
8 Discuss with reference to specific examples the role played by conservation authorities in managing national parks in different countries.
9 Why should the countryside be conserved?
10 How does Wilderness Park management conserve habitats and their associated wildlife?

Multiple-choice questions

Choose the best answer for each of the following questions.

1 The upper boundary of the distribution of life in the atmosphere is about:
    (a) 6.5 kilometres   *
    (b) 10.5 kilometres
    (c) 14.5 kilometres
    (d) 32.5 kilometres

2 The first detectable forms of life appeared during the Archaen which was:
    (a) 4,600 million years ago
    (b) 3,500 million years ago   *
    (c) 460 million years ago
    (d) 350 million years ago

3 The term ecology was first used in 1866 by the biologist:
    (a) Max Ernst
    (b) Charles Darwin
    (c) Ernst Haeckel   *
    (d) Ernst Heinkel

4 Bacteria and fungi belong to which kingdom of living organisms?
    (a) the Funga
    (b) the Plant
    (c) the Protista   *
    (d) the Animal

5 At present how many species on Earth have been described and named?
    (a) 45 thousand
    (b) 950 thousand
    (c) 1.6 million *
    (d) 80 million

6 *The Origin of Means of Natural Selection* was written by:
(a) Alfred Wallace
(b) Charles Darwin   *
(c) Thomas Malthus
(d) Harold Beagle

7 Organisms that have a foreign gene added to them are known as:
(a) morphogenic
(b) bigenic
(c) androgenic
(d) transgenic   *

8 Which of the following means 'ancient life'?
(a) Archaen
(b) Panagean
(c) Proterozoic
(d) Palaeozoic   *

9 The Mesozoic era was dominated by:
(a) fish
(b) reptiles   *
(c) mammals
(d) birds

10 By 1990 how much of the Earth's land surface was classified by UNEP as 'protected sites'?
(a) 0.4 %
(b) 4.4 %   *
(c) 14.4 %
(d) 34 %

11 The world's first national park was:
(a) Yellowstone   *
(b) Cevennes
(c) Snowdonia
(d) Serengeti

12 The US Wilderness Act was passed in:
(a) 1864
(b) 1924
(c) 1944
(d) 1964   *

Figure questions

1 Figure 16.2 illustrates the human activities that cause species extinction. Answer the following questions.
(a) What are the direct human activities responsible for species extinction?
(b) Explain the role of agricultural intensifi-

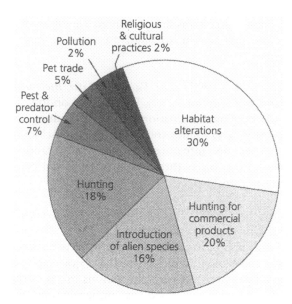

Figure 16.2 Human activities that cause species extinction. Many different factors give rise to the extinction of species, amongst the most important of which are habitat alterations, hunting for commercial products, hunting and the introduction of alien species. More than one activity may be involved in causing the extinction of a particular species. After Figure 14.11 in Cunningham, W.P. and B.W. Saigo (1992) Environmental science: a global concern. *Wm.C. Brown Publishers, Dubuque*

cation in species extinction.
(c) Species extinction is not a new phenomenon, why should we worry about the recent loss of biodiversity?

Answers

(a) Activities that directly target species are hunting for commercial products and sport. At a lower level the pet trade and cultural practices target certain species.

(b) Agricultural intensification has a variety of impacts on biodiversity loss. The alteration and fragmentation of natural areas due to agricultural use destroys habitats directly. The inputs of fertilizers and manure produce a monocultural crop. Chemical treatment, especially pesticides, disrupts the food chain and may poison non-target species.

(c) As humans, our actions are governed by moral law, whereas processes in nature are governed by natural law. Therefore, we have to justify the extinction of species by human action, or inaction, on the basis of these moral principles. This is reinforced by the increased rate of biodiversity loss due to human activities as well as our recent perception of the value of these species and how they may benefit us. The worry, therefore, is a combination of our moral position on their right to live superimposed on the extrinsic value of species.

Short-answer questions

1 Briefly explain the evolution of the biosphere from the Gaia standpoint.

Answer

The biosphere was produced by a complex feedback mechanism whereby biological processes altered the chemistry of the atmosphere by photosynthesis producing an oxygen-rich environment. Organic input into the geochemical cycling process produced the soil layer, whilst the enhanced atmosphere promoted the biological evolution of plants and animals. The soil layer further enhanced this evolution. Surface features and the atmosphere continued to interrelate via weathering processes so that both biotic and abiotic components maintained dynamic stability.

2 What is Biosphere II?

Answer

Biosphere II is an ecological test project started in 1991. Artificial habitats have been created in the controlled environment of a geodesic glass dome. The experiment was designed to investigate the natural cycles and whether they could maintain system stability under experimental conditions.

3 Define ecology.

Answer

Ecology is the study of the relationship between living organisms and their physical environment.

4 Describe the dynamism of biodiversity.

Answer

Biodiversity is not static. It changes continuously as evolution produces new species and some existing species become extinct. This is controlled by a change in ecological conditions, either natural or anthropogenic.

5 List reasons for conserving the Earth's gene pool.

Answer

Retaining the gene pool in a complete state will: improve future agricultural species; improve the characteristics of livestock and domestic animal species; and aid in the development of new medicines.

6 What are mass extinctions?

Answer

Mass extinctions are when a group of organisms become extinct and the lineage nearly always ends without leaving descendants. They may be prompted by significant environmental changes, such as climate change, the development of new competitors, disease, or the loss of food supply. However, many mass extinctions remain enigmas, with much speculation about the exact causal processes.

7 Describe the objectives of the World Conservation Strategy.

Answer

Three key objectives were defined in the World Conservation Strategy. They are all based on sustainable principles of maintaining ecological processes and life-support systems, preserving biodiversity and promoting intergenerational use and maintenance of ecosystems.

8 Outline the use of 'island biogeography'.

Answer

This is the study of plants and animals on isolated islands. Due to their relative lack of interference and in some cases human exploitation, they provide living laboratories for observing evolutionary change. Thus processes producing species diversity, the habitat mosaic, extinction and colonization rates can be monitored.

9 List the key factors in designing nature reserve networks.

Answer

The key factors that need to be taken into account are: the need to conserve as diverse a range of species and habitats as possible; to use available resources in sustainable ways; and to provide a spatial balance between unique sites and those characteristic of the general region.

10 Describe the zoning pattern in biosphere reserves.

Answer

Biosphere reserves consist of three zones: an inner core which acts as a sanctuary against human disturbance; a surrounding buffer zone allowing some activities but essentially protecting the core (both have legal protection); and third, a transition zone where reserve management blends into the general socio-economic development of the region.

Additional references

Erwin, D. E. (1996) 'The mother of mass extinctions'. *Scientific American* 275 (1), 56–62.
A discussion of the likely causes of megafaunal extinctions related to large-scale climatic and tectonic events. Considers the conditions that produced species evolution and adaption and the ability of the Earth to continue supporting life despite global upheaval.

Homewood, B. (1996) 'High and dry in Columbia'. *New Scientist* 150 (2036), 34–7.
Threatened landscape in Columbia is linked to water and energy systems. Human activities pose the threat and conservation policies to counter these via protection are examined.

Lewin, R. (1996) 'A strategy for survival?' *New Scientist* 149 (2017), 14–15.

Ecology and environmental management practices are reviewed against genetic diversity as ways of conserving species. The loss of habitat appears to be the primary factor affecting a species' future.

MacKenzie, D. (1996) 'Seals to the slaughter'. *New Scientist* 149 (2021), 34–9.
A disturbing article that graphically illustrates the threat to seals. Importantly raises the issues of the complexity of food webs in the marine ecosystem. Carrying capacity, variable population sizes and recovery rates are contextualized in the light of limited scientific knowledge.

Middleton, N. (1997) 'Biodiversity'. *Geography Review* 10 (3), 40–1.
The first of three sequential articles on biodiversity (see also below). Explains the concept and its importance to global system function. A very useful box describes taxonomic classifications. Also provides a short summary of the limits of scientific knowledge in quantifying the biotic resource.

Middleton, N. (1997) 'Biodiversity under threat'. *Geography Review* 10 (4), 11–13.
An overview of the main threats to biodiversity. Some useful tables included on extinctions through time and the spatial scale of a variety of habitat loss.

Middleton, N. (1997) 'Conservation of biodiversity'. *Geography Review* 10 (5), 24–5.
A short outline of the main conservation measures applicable to protecting biodiversity.

Tilman, D., Wedin, D. and Knops, J. (1996) 'Productivity and sustainability influenced by biodiversity in grassland ecosystems'. *Nature* 379 (6567), 718–20.
Highlights the importance of biodiversity. Demonstrates that species extinctions threaten ecosystem stability. A clear field-based study supporting interactive, holistic biotic systems.

Web site

www.soton.ac.uk/~engenuir/
A most useful, well-presented introduction to an environmental database. Good coverage of topics related to biodiversity with illustrative case studies.

Aims

- To demonstrate that nature is ordered and explore ways in which this is reflected in the biosphere system.

- To explore the structure and function of terrestrial ecosystems.

- To focus on succession as a mechanism of vegetation development.

- To compare global patterns of vegetation with the ecosystem controls of climate and soils.

- To establish the integrating link between the biotic and abiotic aspects of the environment.

Key-point summary

- Ecosystems are complex, interdependent and organized systems comprising all the living (*biotic*) and non-living (*abiotic*) components of an area. They are of a spatial scale convenient for study and classification. This is usually defined at the local/regional scale. Being open systems they interact with the major environmental systems and as units comprise a mosaic of systems within the biosphere.

- Ecosystems have an ordered structure reflecting their biotic and abiotic components as well as the interaction between them. Organisms may be classified as *producers* or *consumers*, comprising a hierarchical structure of energy flow through the system. This may be described, in relation to feeding habitats, by simple *food chains* and more complex *webs* divided into trophic (nutritional) levels. Classification may also be by numbers at each level (*ecological pyramid*). Population and energy usage underpins the threshold at which the

habitat can holistically operate successfully with individual species operating in specific *ecological riches*. A major threat to this successful operation is the human introduction of non-indigenous species.

- The links to the abiotic components of the ecosystem are via *decomposition* which is vital to major biogeochemical cycles and soil development.

- Ecosystems display the dynamism seen in other systems. Energy flows through the system, being both an input and an output. On the other hand *nutrients* display a feedback mechanism, being recycled and reused. These movements are major influences on the *productivity* of an ecosystem. This may be influenced by *limiting factors* relating to change in other natural systems, e.g. climate, or by human intervention, e.g. soil erosion. Time is an important consideration here, as the equilibrium of the natural ecosystem may respond in an adaptive way to long-term gradual change but break down under short-term significant change.

- The ecosystem concept has merit in facilitating study of the environment. It is of a convenient scale and provides a framework in which to observe and analyse the functioning of nature and natural processes. As a functional unit the effects of human activities can be quantified and appropriate management decisions made. However, there are practical problems associated with the study of ecosystems. Ecosystems are difficult to define spatially as there are few sharp boundaries in nature. Consistency of scale of investigation is therefore problematic. As they are often extremely complex systems, comprehensive analytical measurement is extremely difficult. Human interference has modified ecosystems so that the natural workings are difficult to define. Overall, the concept of natural unity, interaction and *synergy* has made the adoption of an ecosystem approach to environmental management more widespread in the late twentieth century.

- The vegetation community of an ecosystem can be described as part of a sequential development trend (*succession*). The process of *colonization* of unvegetated land and subsequent *seral* stages are stages in succession leading to eventual community equilibrium (*climax community*).

- The end product of succession is *climatic climax vegetation*. As its label suggests it is a product of its climate-dominated environment. Climatic climax vegetation reflects the timescale of succession, being the end-point of the process; though it is argued that this climax is rarely achieved because the environment is rarely stable enough for equilibrium to be achieved.

- *Biomes* are extensive plant and animal communities. The Earth's surface may be divided into a biomic pattern which closely relates to the influence of the climatic system. This pattern reflects altitude as well as latitude.

- *Tundra* and *taiga* have conditions related to cold temperatures. These conditions are conducive to low biodiversity. Tundra is differentiated physically from the taiga by having a *permafrost* zone nearer the surface, and vegetationally by a general absence of trees. Human impacts are related to general settlement infrastructure, mineral extraction and timber removal.

- The *temperate grassland* region consists of *prairie* and *steppe*. They are exploited for grazing in their natural state. Exceeding the carrying capacity results in scrub invasion.

- *Forests* are also found in temperate regions and are widely held to be the climatic climax vegetation. They require deeper soils and more water than grassland; consequently short-term climatic change reducing water supply will reduce tree distribution with a positive feedback to the carbon cycle by removing a major carbon store. The forest structure also reflects its position in relation to land–sea masses and associated climate. Thus, variations of marine west coast forests and humid continental mixed forests exist within the latitudinal band.

- Hot climates establish a variety of biomes, ranging from *desert* through shrub and scrub to *savannah* and forests. They are climatically differentiated by levels of precipitation.

- The most productive biomes are established in the humid tropics (*tropical forests*). Their massive biomass and high biodiversity mean they exert a strong biological control on the Earth. This is demonstrated by their effect on global weather patterns and regional regulation of soil erosion and the hydrological cycle.

Main learning burdles

Food chains and food webs

Some time should be spent on Figures 17.3 and 17.4 so that the dynamics of these hierarchies are understood. This is important if students are to understand some basic ecological dynamics and the applied aspects such as pollution concentration within natural systems.

Nutrient cycling

A review of the system cycles in Chapter 2 should be discussed. The various major environmental systems' involvement should be emphasized to illustrate ecosystem complexity, and how changes in these relationships influence the biosphere.

Key terms

Biomes; climax community; colonization; decomposition; ecological pyramids; ecosystem; food chains and webs; desert; habitat; hot climates; landscape ecology; mangrove; niche; nutrients; optimum; prairie; productivity; restoration ecology; savannah; sere; steppe; succession; taiga; temperate forest; tolerance range; trophic levels; tropical forest; tundra.

Issues for group discussion

Discuss the value of the tropical rainforest

The discussion should focus on framing the issue in relation to its extrinsic worth to natural global system functioning as opposed to its economic worth to the human resource system. The discussion may lead to framing at a variety of scales: local, regional and global. The instructor should encourage the students to discuss the question from another group's perspective, e.g. indigenous rainforest inhabitants; the poor in Brazil; a government with a huge national debt. Both Myers (1988) and Tyler (1990) can usefully provide material at various stages of the discussion.

Discuss how a knowledge of ecosystem structure can aid our management of these systems

This gives the students the chance to analyse key components of the ecosystem. The instructor might prefer to be more prescriptive and give out a list of ecosystem components drawn from the chapter. The discussion should develop from this analytical base to one of synthesis where the students fit the components together in relation to management problems such as the unintentional poisoning of species higher in the food chain.

Selected reading

Brown, D. A. (1993) 'Early nineteenth century grasslands of the Mid-continent Plains'. *Annals of the Association of America Geographers* 83(4), 589–612.
A case study examination of theoretical processes in relation to observed evidence from sediments is examined to suggest historical change in relation to a range of environmental variables.

Myers, N. (1988) 'Tropical forests: Why they matter to us'. *Geography Review* 1(4), 16–19.
The relationship between the tropical forest biome and its biodiversity is discussed in relation to habitat loss. Clear system links are drawn between human consumer patterns and the direct effects on these tropical systems.

Tyler, C. (1990) 'Laying waste'. *Geographical Magazine* 62(1), 26–30.
Presents a clear case of global system interference due to the degradation of tropical rainforests. Physical system interference by the global network of the timber industry is well illustrated and quantified by a series of maps.

Textbooks

Goudie, A. (1993) *The Nature of the Environment*, 3rd edn. Blackwell: Oxford.

Whilst considered as a core physical geography text this book's layout makes it particularly useful in expanding discussion of this chapter. World zones are presented as syntheses of climate, topography and vegetation, reinforcing the zonal pattern presented in this chapter. Furthermore, the many examples of particular ecosystems as components of larger systems provide for case study evaluation.

Reaching, A. J., Thompson, R. D. and Millington, A. C. (1995) *Humid Tropical Environments*. Blackwell: Oxford.
A multidisciplinary perspective draws together the interaction of the environmental systems affecting the humid tropical regions of the world. It is well illustrated with appropriate use of case studies. Environmental hazards and issues related to these environments provide clear evidence of the global importance of their study.

Roberts, N. (ed.) (1994) *The Changing Global Environment*. Blackwell: Oxford.
A range of topical chapters comprise this text. The tropical forest ecosystems are covered and there is a useful section on remote sensing applications.

Viles, H. (ed.) (1988) *Biogeomorphology*. Blackwell: Oxford.
This text offers a comprehensive examination of the major environmental systems and how they interact. Contains sound examinations of the main global cycling mechanisms and surface processes.

Essay questions

1 Explain the factors that influence the spatial distribution of flora and fauna on a global scale.
2 Describe the characteristics of the tundra climate which may pose problems for flora and fauna. How do they adapt?
3 Evaluate the monoclimax theory of succession in the light of case study evidence.
4 Contrast the nutrient cycles of a tropical agricultural system and a tropical rainforest.
5 Discuss the contention that temperature largely controls the altitudinal zonation of vegetation.
6 With reference to one type of ecosystem, examine the nature of human influence on its development.
7 Justify the conservation of ecosystems.
8 'Natural disturbance plays a key role in shaping the structure and function of ecological communities.' Discuss.

9 Critically evaluate ecosystem management in relation to the application of system knowledge.

10 Discuss the arguments that could be used to oppose the removal of large areas of tropical rainforest.

Multiple-choice questions

Choose the best answer for each of the following questions.

1 How much of the solar radiation that reaches the Earth's surface is useful in the process of photosynthesis?
   (a) 23 per cent
   (b) 43 per cent   *
   (c) 62 per cent
   (d) 93 per cent

2 How much of primary productivity becomes available to the top carnivores?
   (a) 1 per cent
   (b) 5 per cent   *
   (c) 50 per cent
   (d) 75 per cent

3 What is the main difference between energy and nutrient flows through ecosystems?
   (a) energy can be reused
   (b) nutrients are lost from ecosystems
   (c) nutrients produce heat
   (d) nutrients can be recycled   *

4 Ecosystem productivity may be influenced by 'limiting factors'. Who established this concept?
   (a) von Leibig   *
   (b) Darwin
   (c) Clements
   (d) Meadows

5 How many successional stages are identified by Clements?
   (a) 5   *
   (b) 3
   (c) 2
   (d) 7

6 High annual precipitation and temperature characterize the:
   (a) temperate rainforest
   (b) taiga
   (c) savannah
   (d) tropical rainforest   *

7 Agenda 21 proposed which one of the following for sustainable forest management:
   (a) use hardwoods for furniture making
   (b) set up small local forest enterprises   *
   (c) use traditional slash and burn practices
   (d) separate trees from urban areas

8 The top layer of tree stratification in the tropical rainforest is characterized by:
   (a) treelets
   (b) lianas
   (c) emergents   *
   (d) the main canopy

9 CITES stands for:
   (a) Committee for International Tropical Eco-system Sites
   (b) Committee for Important Tropical Eco-logical Sites
   (c) Convention on Inter-Tropical Energy Systems
   (d) Convention on International Trade in Endangered Species   *

10 A mangrove swamp is similar to which vegetation in temperate areas?
   (a) saltmarsh   *
   (b) reed beds
   (c) mixed woodland
   (d) peat bog

Figure questions

1 Figure 17.7 represents the energy flows in an ecosystem. Answer the following questions.
   (a) What is another term for the plants component and what is their role?
   (b) Describe the flow of energy.
   (c) Briefly outline how human interference may upset the operation of the ecosystem.

Answers

(a) The plants may also be termed producers. They convert solar energy into chemical form by the process of photosynthesis.

(b) The flow of energy through the system is one-way.

(c) Humans may interfere at various stages within the ecosystem. Air pollution may reduce the amount

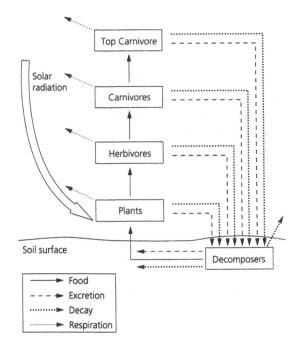

Figure 17.7 Principal energy flows in an ecosystem. Energy in ecosystems is ultimately derived from solar radiation, but it passes through the ecosystem along the food webs. Decomposition is an important part of this process.

of solar radiation reaching the primary producers. Exploitation of minerals will degrade the soil layer and deprive the plants of their decomposed food supply. Pastoral practices may remove the plants for fodder to fuel the livestock industry elsewhere. Humans may also introduce exotic species which may act as predators or competitors, so disrupting the food chain.

2 Figure 17.18 depicts the stratification of the tropical rainforest. Answer the following questions.
  (a) Describe the composition and structure of the tropical rainforest.
  (b) Briefly explain the relationship between climate, soils and vegetation in the tropical rainforest.

Answers

(a) The structure of the tropical rainforest consists of five layers or strata. The top layer consists of tall trees (emergents) which grow through the main canopy layer. Both these are deciduous trees, but due to the continuous growing season they shed

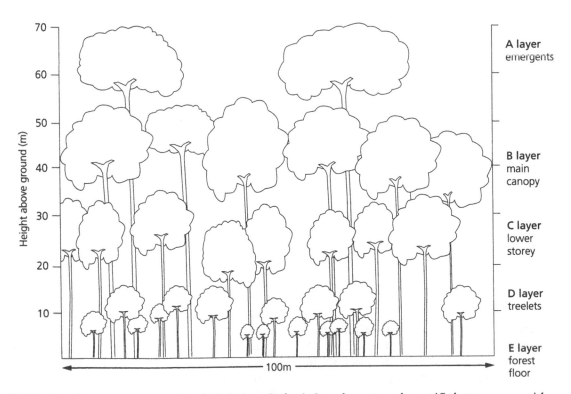

Figure 17.18 Stratification in the rainforest. Typical tropical rainforest has a strongly stratified appearance, with distinctive forms of vegetation at each layer. See text for explanation. After Figure 21.12 in Dury, G.H. (1981) Environmental systems. *Heinemann, London*

their leaves at any time in the year. Below the main canopy is an understorey of trees, analogous with temperate woodland trees, and below these are shrub and forest litter layers. The structure of the top two layers reflects the high photosynthesis rates due to insolation. The species composition below this layer is different due to lower light levels but compared to other biomes it is still highly productive due to the efficient nutrient cycling system.

(b) The rainforest produces a great biomass. This is due to the high solar radiation input, providing energy; heavy precipitation producing a constant moisture surplus; and high temperatures producing a rapid decay of forest floor litter and the recycling of nutrients. Thus climate is optimal for vegetative growth and the production of a fertile soil layer.

Short-answer questions

1 Briefly define the term ecosystem.

Answer

An ecosystem is a system where living organisms and their immediate abiotic environment interact in an interdependent way.

2 What is bioaccumulation?

Answer

Bioaccumulation is the concentration of stable chemical products in species higher up the food chain. Harmful products such as DDT may reach such concentrations that they kill these higher species.

3 Define the term habitat.

Answer

A habitat is the holistic integration of variables that provides the natural home for a plant or animal. A number of species will exist in the same place so, in effect, a habitat supports a community.

4 Describe the ecological effects of limiting factors.

Answer

Limiting factors of sunlight, temperature, nutrient supply and water define the range of tolerance for a particular species. If one or more of these factors changes status then the species becomes less frequent until a tolerance threshold is reached and the species becomes absent.

5 List the five stages of succession.

Answer

The five stages of succession go from the initial creation of bare land (nudation); through seed arrival (migration); plant growth (ecesis); competition (reaction); to achieve an equilibrium condition (stabilization).

6 What are the properties of climax vegetation?

Answer

Climax vegetation is the end-point of succession. Its species composition remains stable through time; it is self-sustaining, maintaining an equilibrium between itself and the soil and climate systems. It has a slow re-establishment rate.

7 Define the term ecotone.

Answer

This is a transitional border zone between biomes. It contains species from both ecological communities. They change position as the areal biomes adjust to environmental change such as climate change.

8 List the major human threats to forests.

Answer

There are four main human threats to forest ecosystems. Firstly, the physical clearance of trees to create land for agriculture. Secondly, the removal of particular species for furniture making and construction. Thirdly, their replacement by plantation silvaculture. And lastly, the impact of air pollutants, especially in the form of acid precipitation.

9 Is the savannah a natural or cultural biome?

Answer

Deliberate burning and grazing of savannah grassland has kept the grass from achieving its maximum height. The debate is whether this human practice has produced savannah from previously cleared tropical forest or whether the climate in this zone produces a climax vegetation of two-metre-high grassland.

10 Describe the reasons why monsoon forests are particularly prone to timber extraction.

Answer

The fewer species associated with the monsoon area mean that commercially it is easier to extract particularly valued species as they are less isolated than in equatorial rainforest. Also, highly valued trees, such as teak, are culturally valued by other South-East Asian countries such as Japan which import them.

Additional references

Holloway, M. (1996) 'Sounding out science'. *Scientific American* 275 (4), 82–8.
An examination of the recovery through time of the ecosystems affected by the *Exxon Valdez* oil spill. Usefully compares intensive small-scale scientific monitoring with a larger-scale system overview.

Holmes, B. and Walker, G. (1996) 'How did paradise begin?' *New Scientist* 151 (2048), 34–7.
Explains the development of the tropical rainforest into the Earth's most complex ecosystem. Clearly links its importance to global biodiversity and outlines threats to its spatial integrity. NB: This is one of the five related articles on tropical rainforests in this issue.

Knight, F. (1996) 'Better than it looks on paper'. *New Scientist* 151 (2049), 16–17.
Considers the whole pattern of world forests in relation to the paper and pulp industry. Discusses whether sustainable forestry can offset pollution from factories. Also examines the benefits or otherwise of plantation silvaculture.

Middleton, N. (1996) 'Sustainability in Finnish forestry'. *Geography Review* 10 (2), 15–17.
The sustainable management of the boreal forest biome is used as a national case study of good practice. Illustrates the consideration of holistic resource use embodying environmental protection and waste management.

Web site

straylight.tamu.edu/bene/bene
The home page of the Biodiversity and Ecosystem Network. It allows access to a range of international projects and is a forum for ideas related to these topics.

Aims

- To examine the nature and development of soils.

- To classify and describe the distribution of soils at a variety of scales.

- To assess the management of soil resources.

Key-point summary

- Soils are comprised of minerals, organic compounds, living organisms, air and water. The combination of these strongly influences the soil type and subsequent resource use, e.g. agriculture. These soil components reflect the links to other major environmental systems and processes. Inputs from near surface geology, biota and the atmosphere influence the soil constituents. Outputs are reflected in the natural productivity of the soil. As vegetation provides both output and input as organic matter it demonstrates a positive feedback loop in the soil system.

- *Soil properties* of *texture*, *structure* and *depth* are spatially variable. They are influenced by other systems, e.g. topography. Soil properties control the store of inputs from other systems, e.g. water, and the release from store as outputs to other systems, e.g. agriculture.

- The basis for soil formation (*pedogenesis*) is the action of weathering processes on surface geology. The speed of this action is largely determined by climate and the susceptibility of the rock to weathering. Through time soils develop a discrete *vertical* profile comprised of *horizons*. These profiles display the action of natural processes, such as *leaching*, through the soil system, as well as allowing a visual classification of different soil types.

- The atmospheric inputs involved in soil profile formation have both physical and chemical effects. In discrete combinations with the soil material they give rise to five major processes: *laterization*, *podsolization*, *calcification*, *acidification* and *salinization*. All these processes may be influenced by human impact, either directly within the soil system or indirectly by effects on other contributing systems.

- At different scales changes in soil type reflect different environmental controls. Climate is the major regulator at the broad regional scale. At the local scale soils will vary with topography and land use. Broadly speaking, *soil classification schemes* reflect these controls and the soil characteristics produced by them.

- Human demands on the soil system are varied and threaten its viability as a productive system. Population growth and poor land management underpin the outcome of soil erosion, physically reducing the system resource, whilst pollution reduces the quality of the resource.

- To counter the human impact *land evaluation* is used as a classificatory and predictive tool in land-use management. The important focus is sustainable management. Here an holistic approach appreciates the influence that climatic vegetation system changes, in concert with increasing human activity, have on the possible degradation of the soil resource. Implementing sustainable management involves *soil conservation* based on the knowledge of how these systems interact and function.

Major learning hurdles

Classification of soils

Soil studies use a large amount of specialized terms

and descriptive classifications that may confuse the students. The instructor should therefore concentrate on the basic processes and restrict soil types to the major sequences related to slopes and specialized weathering regimes, e.g. the tropics. Once the students are conversant with these a closer look at the descriptive changes between similar soil types and what causes them might be appropriately linked to environmental system change.

Key terms

acidification; calcification; carrying capacity; laterization; minerals; organic matter; pedogenesis; pedology; podsolization; soil classification systems; soil conservation; soil erosion; soil mosaics; soil profiles, soil structure; soil textures.

Issues for group discussion

Discuss the issue of whether the quantity or quality of soil available is the most important

The student should read Barraclough (1990) to formulate ideas. A consideration of the uses of soils, extensive or intensive, should be examined before deciding which variable is more appropriate. The instructor should encourage an holistic approach and this should be developed in relation to erosion processes, the loss of soil fertility and damaging inputs. These should be linked to a variety of systems, both natural and human.

Selected reading

Barraclough, D. (1990) 'The Earth taken for granted'. *Geographical Magazine* 62 (3), 36–8.
An appraisal of pressures on the soil resource. Wind and water erosion are discussed in relation to a qualitative viewpoint.

Textbooks

Briggs, D. J. and Courtney, F. M. (1989) *Agriculture and the Environment*. Longman: London.
A range of case studies examines the types of agricultural systems operating globally. These are appraised in relation to their impact on the soil and the biosphere components.

Ellis, S. and Mellor, A. (1995) *Soils and Environment*. Routledge: London.
A broad coverage uses the Earth's soils system as the basis for reviewing a range of environmental topics. Basic soil science is clearly explained and this forms the theoretical basis for evaluating human interactions historically, at present and in the future.

Johnson, D. L. and Lewis, L. A. (1994) *Land Degradation*. Routledge: London.
A readable account of land degradation focusing on biological and geological control as physical determinants of land fragility. Political and economic interference in these areas produce system stress and potential breakdown. Well illustrated with a range of diverse case studies drawn from around the globe.

Troeth, F. R. and Thompson, L. M. (1993) *Soil and Soil Fertility*, 5th edn. Oxford University Press: Oxford.
An excellent introductory text covering the basic soil constituents. Clearly relates soil viability to the water system, agricultural system and biota.

Essay questions

1 Critically evaluate the contention that water is the most important factor in soil development.
2 Examine the catena concept with reference to the spatial variation of soil types.
3 Describe the main profile characteristics of podsolic soils, and on the basis of their main environmental determinants briefly indicate their distribution at the regional scale.
4 Evaluate the factors affecting soil variations with altitude.
5 Discuss the formation and significance of soil structure.
6 Discuss the factors that influence rates of soil erosion on cereal-growing agricultural land.
7 Discuss the causes and consequences of soil degradation in economically less developed countries.
8 'The acidity of stream water is determined by the soil, which in turn is influenced by bedrock and climate.' Discuss.
9 Discuss the importance of soil properties as factors in soil erosion.
10 'Soil erosion is largely controlled by the type of soil management techniques employed in an area.' Evaluate.

Multiple-choice questions

Choose the best answer for each of the following questions.

1 Calcification produces:
   (a) neutral or basic soils   *
   (b) acid soils
   (c) waterlogged soils
   (d) heavy soils

2 Most soils can be assigned to three broad categories. Which one is not one of these categories?
   (a) zonal
   (b) intrazonal
   (c) interzonal   *
   (d) azonal

3 Soil conservation programmes were first instigated in the USA in the:
   (a) 1870s
   (b) 1910s
   (c) 1930s   *
   (d) 1950s

4 Which one of the following is true:
   (a) fine sand is coarser than silt   *
   (b) clay is coarser than silt
   (c) very fine sand is finer than silt
   (d) gravel is finer than coarse sand

5 Soil structure classes can be depicted in a diagram shaped like a:
   (a) square
   (b) triangle   *
   (c) circle
   (d) diamond

6 Latosols have an insoluble layer in the:
   (a) C-horizon
   (b) R-horizon
   (c) A-horizon   *
   (d) B-horizon

7 The smallest three-dimensional unit of soil, defining a block of relative uniform properties is called a:
   (a) mosaic
   (b) pedon   *
   (c) catena
   (d) profile

8 Aluminium and iron accumulate mainly in soils of:
   (a) humid climates   *
   (b) cold climates
   (c) dry climates
   (d) hot climates

9 Rain splash may detach and move particles of fine sand and silt up to a distance of:
   (a) 15 metres
   (b) 1.5 metres   *
   (c) 50 centimetres
   (d) 15 centimetres

10 Loess is a:
   (a) province in China
   (b) yellow estuarine deposit
   (c) ploughing technique
   (d) fine-grained wind-deposited loam   *

Figure question

1 Figure 18.6 illustrates the characteristics of a typical soil profile. Answer the following questions.
   (a) Name and describe the three main processes operating in the soil profile.
   (b) What is the effect of these processes?

Answers

(a) The movement of water downward through the soil is by the process of leaching. Material is moved through the soil zone by eluviation and the precipitation and accumulation lower down in the B-horizon is the process of illuviation.

(b) Eluviation results in the depletion of soil materials in the upper horizons, whilst leaching removes water-soluble minerals from the upper horizons. Both contribute to accumulation in the B-horizon (illuviation) whilst leached minerals may enter the groundwater zone.

Short-answer questions

1 List the main types of parent material promoting soil development.

Answer

Soil may develop from weathered rock; volcanic ash; glacial deposits; floodplain sediments; stabilized dunes; and newly exposed coastlines.

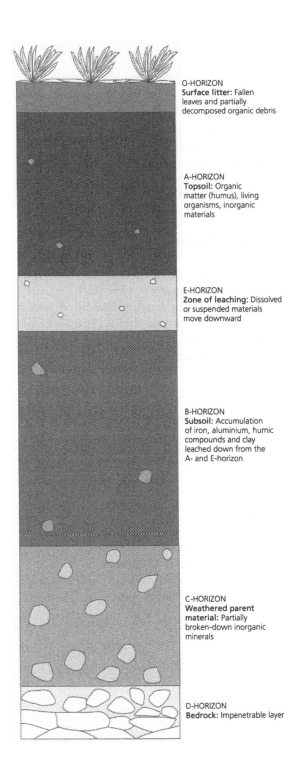

O-HORIZON
Surface litter: Fallen leaves and partially decomposed organic debris

A-HORIZON
Topsoil: Organic matter (humus), living organisms, inorganic materials

E-HORIZON
Zone of leaching: Dissolved or suspended materials move downward

B-HORIZON
Subsoil: Accumulation of iron, aluminium, humic compounds and clay leached down from the A- and E-horizon

C-HORIZON
Weathered parent material: Partially broken-down inorganic minerals

D-HORIZON
Bedrock: Impenetrable layer

Figure 18.6 Characteristics of a typical soil profile. Most soil profiles display evidence of a number of different horizons, although the number, thickness and character of each horizon varies between different soil types. After Figure 11.5 in Cunningham, W. P. and B. W. Saigo (1992) Environmental science: a global concern. *Wm.C. Brown Publishers, Dubuque*

2 What factors control soil depth?

Answer

A range of factors controls the depth of soil. Depth usually increases through time due to bedrock weathering and organic accumulation. To a large extent these are controlled by climate. Topography influences soil depth with runoff and mass movement removing soil from steeper slopes and causing it to accumulate on gentler-sloping downslope facets. Finally a wide range of human activities reduces soil depth by either compaction or loss (soil erosion).

3 Define the term pedogenesis.

Answer

Pedogenesis is the formation and development of a soil. It involves the combined effect of climatic, topographical, geological and biotic processes.

4 List the main chemical processes involved in the formation of a soil.

Answer

The main chemical processes are: hydrolysis; hydration; oxidation; and reduction.

5 Describe the concept of a catena.

Answer

The interactions of the various factors involved in soil formation produce a mosaic of soils. The controls of slope and associated drainage have very strong influences on this pattern. Downslope change often produces a clearly defined sequence of soil profile changes and this is referred to as a catena. The catena concept demonstrates the importance of water in soil formation.

6 What are pedalfers and pedocals?

Answer

These are soils which have concentrations of

particular elements. Pedalfers accumulate aluminium and iron, important for laterization and podsolization. Pedocals accumulate calcium and are common in less humid, drier climates than those which promote pedalfers.

7 Differentiate between soil erosion and land degradation.

Answer

Soil erosion refers to the physical removal of topsoil. Land degradation is the reduction in the quality and usefulness of the soil whilst it remains in situ.

8 List the main human activities promoting soil erosion.

Answer

The main human factors influencing soil erosion are: activities that involve removing the vegetation cover, e.g. deforestation and overgrazing; in previously cleared areas poor agricultural practices such as cultivating slopes without terracing and winter ploughing.

9 List the ways soil erosion by water may be managed.

Answer

The prevention or reduction of water as a soil-removing agent can be achieved by maintaining a vegetative cover throughout the year; by contour ploughing; terracing; and physical restraint such as netting.

10 Outline means of controlling wind erosion.

Answer

The maintenance of a plant cover reduces wind velocity and the energy for loose material removal. More spatially designed systems involve a variety of windbreak mechanisms, either formal, permanent shelterbelts or intercropping rows of tall plants.

Additional references

Higgitt, D. (1997) 'Underground story: Local variation in soils'. *Geography Review* 10 (4), 19–21.
A basic introduction to the soil classification system. Well illustrated and outlines the major factors influencing soil formation. Useful local-scale catena case study represented.

Tegen, I., Lacis, A. A. and Fung, I. (1996) 'The influence on climate forcing of mineral aerosols from disturbed soils'. *Nature* 380 (6568), 419–22.
Human land-use practices that disturb soils are seen as an input to the atmospheric system. Increased heating modifies the dynamics of the atmosphere.

Web site

www.ncg.nrcs.gov/welcome
This is the US Natural Resources Conservation Service page. It allows access to topical links, including soil, vegetation and water.

Aims

- To outline the implications for the Earth of the complex interdependencies between societies and the environment.

- To introduce a range of methods to aid the formulation of environmental management policy.

Key-point summary

- The *United Nations Conference on Environment and Development* (1992) was a milestone in global environmental appreciation. Environmental problems were put on the political stage and future policy expressed under *Agenda 21*. However, the conference highlighted the development chasm between the *North* and *South* with each having different priorities for future resource management. Global solutions to global problems were emphasized and mutual compromise seen as the way forward, e.g. debt for nature agreements.

- *Sustainable development* is viewed as the framework for viable long-term environmental management. An appreciation of how the Earth's systems interact requires the concept to be applied at all scales from the local to global.

- Recently a number of new evaluative tools and methods have been developed to assess human impact on environmental systems. These all provide evidence for feeding into the policy-making framework of environmental management. Key ones are:

  1 *environmental impact assessment* which attempts to predict likely changes in the environmental systems of a discrete area due to a particular human activity, e.g. a new industrial plant. In certain countries this is a statutory requirement;

  2 *environmental modelling* reduces the complexity of natural environmental systems and their interactions with other systems. The key workings of the model may then be assessed under a variety of input, flow and store conditions enabling optimum levels related to a sustainable output to be predicted. This information may then feed into the policy-making process;

  3 an *environmental audit* which provides a baseline inventory of resource use, practices and procedures that affect the environment. It allows a systematic evaluation of our use of resources, allowing appraisal of good and bad practice in relation to sustainable management;

  4 *environmental databases* provide contemporary records of the environmental resource base at a variety of scales. The increasing capacity of *information technology* allows this data to be available to a much greater range of people involved in environmental decision making;

  5 at the national level *state-of-the-environment reports* provide contemporary and retrospective overviews of countries' development programmes and their impact on the environment;

  6 spatial analysis of the environment can be provided by *geographical information systems*. This application of computer technology provides mapping, analysis and predictive modelling from environmental databases at a variety of spatial scales, comprising many variables;

  7 *ecosystems management* provides an integrating, holistic view of system interactions. It allows an appreciation of the complexity of the

Earth's functions and has an application as the basic principle for guiding sustainable management options;

- Scientific knowledge of the Earth has grown rapidly over the last 20–30 years, utilizing increasingly sophisticated monitoring techniques and technologies. However, as our knowledge of the Earth's attributes and functions increases we continue to unveil increasingly complex systems that pose as many new questions as they answer old ones.

Key terms

Geographical information system; monitoring systems; scientific evidence and understanding; sustainability; synthesis; systems.

Issues for group discussion

As this is the concluding chapter, the themes of the book should be brought together. A useful way is to use specific case studies of environmental problems and their management and discuss them in the context of the key terms above. A logical sequence of questions is:

1   What is the problem and what are its effects now and in the future?
2   What do we know about how the problem operates from our monitoring of it and knowledge of the Earth?
3   How do we manage it and how certain are we of the results?

The first two questions relate very much to multidisciplinary evidence and techniques. The last question is much more subjective and reflects the variabilities of the policy-making process. Hemming (1992) is essential reading to emphasize that, without human agreement and the will for action, the environment is gradually becoming more and more at risk. A no less important conclusion is that interest groups can utilize the same half-knowledge and trans-science to produce arguments to support their own diverse agendas.

Selected reading

Hemming, J. (1992) 'Reactions to Rio'. *Geographical Magazine* 64 (9), 23–5.

A brief overview of the problems of producing international environmental policy. Clearly illustrates the differences between media rhetoric and marketing façade and the reality of concrete outcomes.

Texts

Adams, W. M. (1990) *Green Development: Environment and Sustainability in the Third World.* Routledge: London.
This text contains useful material on environmental impact assessment. Good counterpoint to studies from economically developed nations.

Baarschers, W. H. (1996) *Eco-facts and Eco-fiction: Understanding the Environment Debate.* Routledge: London.
A good discussional text that develops course themes related to scientific knowledge and societal truths about the environment. Useful in its balanced approach to the presentation of evidence so that the student may have a more holistic impression of the environment.

Barde, J. P. and Pearce, D. W. (eds) (1991) *Valuing the Environment.* Earthscan: London.
Presents a range of methodologies, introduced in this chapter, for appraising the environment. Useful in outlining techniques in practical sessions.

Barton, H. and Bruder, N. (1995) *A Guide to Local Environmental Activity.* Earthscan: London.
Provides examples of how to conduct a range of environmental audits. Importantly the text emphasizes spatial interdependencies at a variety of scales from the local to global. Usefully fosters a deep consideration of actions from place to place.

Berkes, F. (ed.) (1989) *Common Property Resources: Ecology and Community-based Sustainable Development.* Belhaven: London.
Contains useful discussional material in relation to the future use of the commons. Fosters and re-emphasizes a global approach to protecting our biosphere.

Fischer, F. and Black, M. (eds) (1995) *Greening Environmental Policy: The Politics of a Sustainable Future.* Paul Chapman: London.
This book brings out the importance of the effect of environmental degradation in undermining economic and political systems. National- and international-scale management is clearly identified

in a series of case study papers from around the world. Provides a sound progression from philosophy through policy to practice.

Gilpin, A. (1995) *Environmental Impact Assessment: Cutting Edge for the Twenty First Century.* Cambridge University Press: Cambridge.
A thorough up-to-date examination of the procedures and applications of EIAs for contemporary and future environmental management.

Goodchild, M. F., Parks, B. O. and Steyaert, L. T. (eds) (1993) *Environmental Modelling with GIS.* Oxford University Press: Oxford.
A comprehensive, wide-ranging overview of GIS applied to environmental systems. Predictions of response to change with management implications are discussed. Applications would illustrate most topical chapters in *The Environment*.

Norton, B. G. (1991) *Towards Unity among Environmentalists.* Oxford University Press: Oxford.
The author outlines a variety of environmental philosophies and relates them to applied use in a variety of ecosystems. A useful book to provoke discussion amongst students.

Essay questions

1 How optimistic are you about the possibility of achieving a sustainable future?
2 Discuss the view that, despite continuing uncertainty, natural science remains a vital tool for environmental policy makers.
3 Identify the major factors to be considered and describe the methods you would use to carry out an environmental impact assessment on a major development of your choice.
4 Will finite resources or pollution act as natural limits to growth?
5 Critically examine the processes through which pollution has become globalized.
6 Critically discuss how the time dimension affects the perception and management of environmental hazards.
7 'The long-term management of resources conflicts with short-term exploitation.' Discuss.
8 'Environmental problems are a problem of society.' Evaluate this view.
9 How realistic a resource are the 'alternative' sources of energy?
10 Assess the role of remote sensing in environmental management.

Additional references

Rice, R. E., Gullison, R. E. and Reid, J. W. (1997) 'Can sustainable management save tropical forests?'. *Scientific American* 276 (4), 34–9.
This is a rigorous debate concerning the effectiveness of sustainable management of tropical forests. Issues are raised in relation to the conflicts of interest between economics, politics and conservation. An holistic, pragmatic review suggests possible solutions to prevent further degradation of the tropical forest biome.

Web site

www.geo.ed.ac.uk/home/gishome
A University of Edinburgh site that facilitates access to a range of GIS references and resources. It is applicable to the range of methodologies outlined in this chapter.

The topics and issues presented in *The Environment* lend themselves to alternative methods of assessed assignments. These are complementary to the more traditional approaches laid out in the rest of this manual. Different students will have different mixes of skills and abilities and a range of assessment items both gives them the opportunities to demonstrate these and provides valuable practice in transferable skills applicable to the workplace. The majority of the assignments that follow act as team-building exercises but may be produced individually if required.

Environmental audits

These are useful as they allow students to relate to good environmental practice in organizations. They draw on and consolidate a major theme in *The Environment* by investigating a range of interactions between the human resource use system and its impact on natural environmental systems. The following subsections relate to how the assessment may be presented to the students for assessment by the instructor.

Content and structure of the environmental audit

The environmental audit will require you to utilize skills. You will need to obtain background secondary information, draw up an action plan, implement your plan to obtain primary data from which to produce your audit.

The environmental audit assessment may be divided into the following sections:

1   A brief introduction to the environment and the concept of sustainable development, illustrating why these are issues of central importance and how they are relevant to organizations and, in particular, the organization/operation being audited by you.

2   An examination of environmental management systems (a structured framework of environmental good practice relevant to organizational operation), including concise details of the relevant national/international legislation applicable to the organization being audited, highlighting the key objectives and requirements for environmental accreditation.

3   The specific audit carried out by you in relation to a topic area (or areas) related to the relevant environmental legislation.

4   A list of key recommendations from your audit in relation to:
   (a) reducing environmental impact;
   (b) implementing or adapting an existing environmental management system;
   (c) complying with relevant legislation.

5   A summary of the environmental audit experience as a work task in relation to skills and knowledge used and gained.

Overview

The attainment of environmental objectives within an organization requires an environmental management system (EMS). To integrate environmental considerations into an organization's practice it is necessary, therefore, to implement an EMS. An EMS is a system which checks the policies and practices of an organization and how these may be improved to refine their environmental performance. British Standard 7750 was the world's first standard for environmental management systems, and whilst rather outdated may be considered as a blueprint for any EMS. An environmental audit may be considered as the initial stage of information gathering for the instigation of this system or to appraise a system already in place. The environmental audit produced by you should concentrate on providing a register of environmental effects in relation to the topic(s) being audited, e.g. waste management, energy usage.

Thus, your audit should provide a systematic evaluation of the effectiveness of the organization and its performance in fulfilling sustainable environmental policies and objectives. An exemplar of the structure for sections 3 and 4 may be:

(a) an environmental effects evaluation in terms of one or more of the following variables:
 – water, waste, energy, transport, air, packaging;
(b) an environmental effects register resulting from the above evaluation;
(c) an appraisal of environmental management in terms of:
 – staff awareness
 – command structure relaying environmental information;
(d) recommendations on how environmental impact can be reduced;
(e) recommendations for setting up or amending an environmental management system.

Environmental impact assessment (EIA)

An EIA is an objective appraisal of the potential impacts of a major development on the physical environment. Important aspects of the human environment may also be considered. The focus is on identifying key disturbances prior to development in order to keep environmental disruption to a minimum.

The existing situation is described by inventory, and future change, without the development, assessed. The proposed development is then outlined and the probable environmental impact recorded on a matrix, see Table 20.1. Differences between the change without the development and those produced by the matrix may now be identified. Proposals to reduce these identified impacts should then be discussed, focusing on

positive modifications to reduce environmental impact and unavoidable major adverse impacts. A conclusion regarding the veracity of the development should be drawn.

This type of exercise is suitable for either individual or group work. A useful test of the application of an EIA is to use an undeveloped area of land for the EIA and then, after completing the exercise, to take the students to a reasonably adjacent area of similar landscape where the assessed type of development has already been in place, e.g. coastal defence works. This allows the students' findings from the matrix to be compared in general terms to what would actually happen.

Media file

The production of a contemporary media record will allow the connection of topics and issues raised in the course text to be related to day-to-day experience. In addition it will stimulate interest beyond the course timetable and provide ongoing applied revision.

The key to a good media file is initial planning at the start of the course. The students should be made aware of the total course content via chapter and section headings. Special mention should be made of the sections embodying environmental problems and given the breadth of the text this may be the best focus for the media file. Whilst this could be done on an individual basis, it is possible for the group to work as a team. In this instance chapters may be allotted to course individuals or small groups and the whole group builds up total resource for use by the course members. Resources may be gathered from local and national press to illustrate the variety of scales and views of the environment. To build up additional search skills the use of newspaper CD-ROMS should be encouraged, but, as with the use of the Internet, selectivity must be demanded.

Poster presentation

Poster presentations are useful assignments as they are extensively used in a variety of workplace situations for communicating information and ideas.

A poster is basically a panel used to display a thematic study. Student posters will normally consist of a number of illustrations, e.g. maps, diagrams, etc., linked by explanatory text. The whole should be a unified coherent presentation requiring no further explanation.

Table 20.1 Environmental Impact Assessment (Leopold matrix)

	Existing conditions of the environment																			
	PHYSICAL CHARACTERISTICS							FLORA				FAUNA								
Types of change	Landform	Soils	Erosion	Deposition	Stability	Water flow	Air movements	Trees/Shrubs	Grasses	Microflora	Natural barriers	Birds	Animals	Insects	Microfauna	Scenic quality	Access	Population density	Man-made barriers	Totals
	1	2	3	4	5	6	7	8	9	10	11	12	13	14	15	16	17	18	19	
Beach change																				
Ground cover change																				
Habitat change																				
Surface change																				
Barriers and fences																				
Landscaping																				
Human movement																				
Access changes																				
Physical appearance																				
Totals																				

Complete the matrix as follows:

1 Place a diagonal line in each box where you think an impact will occur.
2 (a) Score above the line the size of the interaction (i.e. large or small scale).
   (b) Score below the line an estimate of the importance of the interaction (i.e. how damaging the impact might be).
   In each case the score should be between 1 and 4 (1 = low, 4 = high).
3 Total the row and column scores.
4 The total effect of the parts of the scheme (rows) and the potential impact on each part of the environment (columns) can be determined and interpreted.

From the outset students should be made aware of the opportunities provided by this type of assignment, such as:

(a) it encourages selectivity in presenting the key-points from a range of complex information, issues and ideas;

(b) it encourages synthesis by selecting appropriate combinations of illustration and text to provide a coherent display;

(c) it encourages teamwork and allows self-assessment of individuals within the group to determine the attributes they can give to the team, e.g. computer graphics skills;

(d) the assessment by an audience of staff and peers produces a greater variety of critical feedback than more traditional methods.

It may be appropriate to use staff from another department, e.g. Art and Design or Marketing to give a short illustrative talk on display techniques, such as titling, colour, layout and flow.

An example of an assignment topic and assessment pro-forma is given below.

Poster presentation assessment item

The poster presentation is a visual display of both text and illustration. It should be able to be examined as a complete piece of work that is self-explanatory. However, the group should provide some verbal support if questioned by the assessing group. The poster topic is Water Pollution and Health. Each group is required to choose a case study topic or area at an appropriate scale.

The poster presentation should meet the following criteria for assessment:

1 a theoretical basis explaining how the pollution is caused;

2 a full explanation of the associated health problems including:
– who is affected/at risk
– the pathways through which pollution reaches the target population
– strategies employed to alleviate the problem, e.g. legislation, technological, health promotion;

3 the above should provide a logical and coherent whole;

4 illustrate a degree of presentational creativity without compromising content;

5 ensure that the visual and verbal communication is clear and convincing;

6 ensure that there is a clear indication of teamwork.

The above criteria have equal weighting and will be assessed by peer and tutor (50/50).

You will have 30 minutes at the start of the session to put up your poster.

Role-play exercise

This allows students to enact the decision-making process. It facilitates the framing of an environmental issue from a variety of perspectives, allowing a range of views to be expressed and a decision on policy to be reached via compromise.

To assign roles the environmental issue needs to be broken down into its elements (usefully, systems involved in the issue) and also parties likely to have a vested interest in the issue. Therefore, the decision-making team may comprise a series of system experts (e.g. ecologist, geologist, climatologist, hydrologist, geomorphologist) and a group of interested parties, such as politicians (local, national or international), economists, multinational company representatives and non-governmental organizations.

Dependent on group numbers, a variety of situations may be tried. A useful one is for the interested parties to prepare their arguments and call on evidence from selected systems experts. With a large group, decision-making role-play may be used frequently and therefore students not assigned roles for a particular issue provide a voting audience. In smaller groups the instructor or an individual student may act as a decision-making chairperson. To provide a link between the policy discussion group and the voting audience it is useful to allocate a small group of students as media reporters who produce updates of discussion and summarize the main issues. These may be disseminated to the audience for information.

Table 20.2 *Peer appraisal form: presentation*

Title:
Course Number:
Students:

A SUBJECT MATTER

1	Theoretical basis Clear use of theory/concepts	5 4 3 2 1 0	No use of theory/concepts

Comments

2	Contents Full explanation of the topic	5 4 3 2 1 0	Inadequate explanation

Comments

3	Organization Subject matter is well integrated	5 4 3 2 1 0	Subject matter is disorganized and muddled

Comments

B PRESENTATION

4	Creativity Wide range of approaches, imaginative, interesting	5 4 3 2 1 0	Narrow range of approaches, dull, predictable

Comments

5	Communication Clear, well presented, convincing	5 4 3 2 1 0	Unclear, poorly presented, unconvincing

Comments

6	Teamwork Co-operative involvement of full team clearly evident	5 4 3 2 1 0	No co-operation, looks like one person's work

Comments

Score (out of 30)

Printed and bound by CPI Group (UK) Ltd, Croydon, CR0 4YY

21/10/2024

01777093-0017